Lecture Notes in Computer Science 8364

Commenced Publication in 1973
Founding and Former Series Editors:
Gerhard Goos, Juris Hartmanis, and Jan van Leeuwen

Jan Jürjens   Frank Piessens
Nataliia Bielova (Eds.)

Engineering Secure Software and Systems

6th International Symposium, ESSoS 2014
Munich, Germany, February 26-28, 2014
Proceedings

 Springer

Volume Editors

Jan Jürjens
Technical University Dortmund
Department of Computer Science
Dortmund, Germany
E-mail: jan.juerjens@isst.fraunhofer.de

Frank Piessens
KU Leuven
Department of Computer Science
Heverlee, Belgium
E-mail: frank.piessens@cs.kuleuven.be

Nataliia Bielova
Inria Sophia Antipolis – Mediterranee
Sophia Antipolis Cedex, France
E-mail: nataliia.bielova@inria.fr

ISSN 0302-9743   e-ISSN 1611-3349
ISBN 978-3-319-04896-3   e-ISBN 978-3-319-04897-0
DOI 10.1007/978-3-319-04897-0
Springer Cham Heidelberg New York Dordrecht London

Library of Congress Control Number: 2014930756

CR Subject Classification (1998): E.3, D.4.6, D.2.1, D.2.4, F.3.1, K.6.5

LNCS Sublibrary: SL 4 – Security and Cryptology

Typesetting: Camera-ready by author, data conversion by Scientific Publishing Services, Chennai, India

Printed on acid-free paper

Springer is part of Springer Science+Business Media (www.springer.com)

Preface

It is our pleasure to welcome you to the 6th International Symposium on Engineering Secure Software and Systems (ESSoS 2014). This event in a maturing series of symposia attempts to bridge the gap between the scientific communities from software engineering and security with the goal of supporting secure software development. The parallel technical sponsorship from ACM SIGSAC (the ACM interest group in security) and ACM SIGSOFT (the ACM interest group in software engineering) demonstrates the support from both communities and the need for providing such a bridge.

Security mechanisms and the act of software development usually go hand in hand. It is generally not enough to ensure correct functioning of the security mechanisms used. They cannot be "blindly" inserted into a security-critical system, but the overall system development must take security aspects into account in a coherent way. Building trustworthy components does not suffice, since the interconnections and interactions of components play a significant role in trustworthiness. Lastly, while functional requirements are generally analyzed carefully in systems development, security considerations often arise after the fact. Adding security as an afterthought, however, often leads to problems. Ad hoc development can lead to the deployment of systems that do not satisfy important security requirements. Thus, a sound methodology supporting secure systems development is needed. The presentations and associated publications at ESSoS 2014 contribute to this goal in several directions: On the one hand, with secure software engineering results for specific application domains (such as Web and mobile security). On the other hand, improving specific methods in secure software engineering (such as model-based security or formal methods). A third set of presentations presents real-life applications of secure software engineering approaches.

The conference program featured three major keynotes from Ross Anderson (University of Cambridge) on the psychology of security, Adrian Perrig (ETH Zurich) on scalability, control, and isolation for next-generation networks, and Stephan Micklitz (Google Munich) on human factors and strong authentication, as well as a set of research and idea papers. In response to the call for papers, 55 papers were submitted. The Program Committee selected 11 full-paper contributions (20%), presenting new research results on engineering secure software and systems. In addition, there are four idea papers, giving a concise account of new ideas in the early stages of research.

Many individuals and organizations have contributed to the success of this event. First of all, we would like to express our appreciation to the authors of the submitted papers and to the Program Committee members and external referees, who provided timely and relevant reviews. Many thanks go to the Steering Committee for supporting this series of symposia, and to all the members of the Organizing Committee for their tremendous work and for excelling in their respective tasks. The DistriNet research group of the KU Leuven did an excellent job with the website and the advertising for the conference. Finally, we owe gratitude to ACM SIGSAC/SIGSOFT, IEEE TCSP, and LNCS for continuing to support us in this series of symposia.

December 2013

Jan Jürjens
Frank Piessens
Nataliia Bielova

Conference Organization

General Chair

Alexander Pretschner   Technische Universität München, Germany

Program Co-chairs

Jan Jürjens   TU Dortmund and Fraunhofer ISST, Germany
Frank Piessens   Katholieke Universiteit Leuven, Belgium

Publication Chair

Nataliia Bielova   Inria Sophia Antipolis, France

Publicity Chair

Pieter Philippaerts   Katholieke Universiteit Leuven, Belgium

Web Chair

Ghita Saevels   Katholieke Universiteit Leuven, Belgium

Local Arrangements Chair

Regina Jourdan   Technische Universität München, Germany

Steering Committee

Jorge Cuellar   Siemens AG, Germany
Wouter Joosen   Katholieke Universiteit Leuven, Belgium
Fabio Massacci   Universitá di Trento, Italy
Gary McGraw   Cigital, USA
Bashar Nuseibeh   The Open University, UK
Daniel Wallach   Rice University, USA

Program Committee

Ruth Breu	University of Innsbruck, Austria
Lorenzo Cavallaro	Royal Holloway, University of London, UK
Anupam Datta	Carnegie Mellon University, USA
Werner Dietl	University of Washington, USA
François Dupressoir	IMDEA, Spain
Eduardo Fernandez	Florida Atlantic University, USA
Eduardo Fernandez-Medina Paton	Universidad de Castilla-La Mancha, Spain
Cormac Flanagan	U.C. Santa Cruz, USA
Dieter Gollmann	TU Hamburg-Harburg, Germany
Arjun Guha	Cornell University, USA
Christian Hammer	Saarland University, Germany
Hannes Hartenstein	Karlsruher Institut für Technologie, Germany
Maritta Heisel	University of Duisburg Essen, Germany
Peter Herrmann	NTNU, Trondheim, Norway
Valerie Issarny	Inria, France
Limin Jia	Carnegie Mellon University, USA
Martin Johns	SAP Research, Germany
Jay Ligatti	University of South Florida, USA
Heiko Mantel	TU Darmstadt, Germany
Haris Mouratidis	University of East London, UK
Martín Ochoa	Siemens AG, Germany
Jae Park	University of Texas at San Antonio, USA
Erik Poll	RU Nijmegen, The Netherlands
Wolfgang Reif	University of Augsburg, Germany
Riccardo Scandariato	Katholieke Universiteit Leuven, Belgium
Ketil Stølen	SINTEF, Norway
Steve Zdancewic	University of Pennsylvania, USA
Mohammad Zulkernine	Queens University, Canada

Additional Reviewers

Azadeh Alebrahim	Kuzman Katkalov	David Pfaff
Kristian Beckers	Basel Katt	Fredrik Seehusen
Abhishek Bichhawat	Johannes Leupolz	Christian Sillaber
Marian Borek	Yan Li	Bjørnar Solhaug
Michael Brunner	Steffen Lortz	Barbara Sprick
Gencer Erdogan	Rene Meis	Kurt Stenzel
Stephan Faßbender	Jan Tobias Muehlberg	Lianshan Sun
Matthias Gander	Sebastian Pape	Marie Walter
Jinwei Hu	Davide Papini	Philipp Zech

Sponsoring Institutions

Technische Universität München, Germany

NESSoS FP7 Project, Network of Excellence on Engineering Secure Future Internet Software Services and Systems, www.nessos-project.eu

Keynote Abstracts

The Psychology of Security

Ross Anderson

University of Cambridge, UK

Abstract. A fascinating dialogue is developing between psychologists and security engineers. At the macro scale, societal overreactions to terrorism are founded on the misperception of risk and uncertainty, which has deep psychological roots. At the micro scale, more and more crimes involve deception; as security engineering gets better, it's easier to mislead people than to hack computers or hack through walls. Many frauds can be explained in terms of the heuristics and biases that we have retained from our ancestral evolutionary environment.

At an even deeper level, the psychology of security touches on fundamental scientific and philosophical problems. The 'Machiavellian Brain' hypothesis states that we evolved high intelligence not to make better tools, but to use other monkeys better as tools: primates who were better at deception, or at detecting deception in others, left more descendants. Yet the move online is changing the parameters of deception, and robbing us of many of the signals we use to make trust judgments in the "real" world; it's a lot easier to copy a bank website than it is to copy a bank. Many systems fail because the security usability has not been thought through: the designers have different mental models of threats and protection mechanisms from users. And misperceptions cause security markets to fail: many users buy snake oil, while others distrust quite serviceable mechanisms.

Security is both a feeling and a reality, and they're different. The gap gets ever wider, and ever more important. In this talk I will describe the rapidly-growing field of security psychology which is bringing together security engineers not just with psychologists but with behavioural economists, anthropologists and even philosophers to develop new approaches to risk, fraud and deception in the complex socio-technical systems on which we are all coming to rely.

SCION: Scalability, Control, and Isolation On Next-Generation Networks

Adrian Perrig

Swiss Federal Institute of Technology (ETH), Switherland

Abstract. We present an Internet architecture designed to provide route control, failure isolation, and explicit trust information for end-to-end communications. SCION separates ASes into groups of independent routing sub-planes, called isolation domains, which then interconnect to form complete routes. Isolation domains provide natural separation of routing failures and human misconfiguration, give endpoints strong control for both inbound and outbound traffic, provide meaningful and enforceable trust, and enable scalable routing updates with high path freshness. As a result, our architecture provides strong resilience and security properties as an intrinsic consequence of good design principles, avoiding piecemeal add-on protocols as security patches. Meanwhile, SCION only assumes that a few top-tier ISPs in the isolation domain are trusted for providing reliable end-to-end communications, thus achieving a small Trusted Computing Base. Both our security analysis and evaluation results show that SCION naturally prevents numerous attacks and provides a high level of resilience, scalability, control, and isolation.

Human Factors and Strong Authentication

Stephan Micklitz

Google Munich, Germany

Abstract. Google's login team began focusing on strong authentication in the spring of 2008, and in almost six years we have come a long way in protecting our users. In this presentation we will talk about the progress we have made since then, such as introducing strict 2-step verification, risk-based login challenges and OpenID-style login.

We will then identify the biggest challenges we are currently facing in establishing stronger authentication – both from a technological as well as a usability point of view. We will also talk important privacy considerations for such systems, and how we are addressing them. Next we will look into our plan to address these challenges in the next years ahead of us, making use of technological developments, e.g. the vastly increased adoption of smart mobile devices.

Table of Contents

Model-Based Security

Formal Methods

Web and Mobile Security

Applications

Detecting Code Reuse Attacks with a Model of Conformant Program Execution

Emily R. Jacobson, Andrew R. Bernat,
William R. Williams, and Barton P. Miller

Computer Sciences Department, University of Wisconsin
{jacobson,bernat,bill,bart}@cs.wisc.edu

Abstract. Code reuse attacks circumvent traditional program protection mechanisms such as $W \oplus X$ by constructing exploits from code already present within a process. Existing techniques to defend against these attacks provide ad hoc solutions or lack in features necessary to provide comprehensive and adoptable solutions. We present a systematic approach based on first principles for the efficient, robust detection of these attacks; our work enforces expected program behavior instead of defending against anticipated attacks. We define *conformant program execution* ($\mathcal{CPE}$) as a set of requirements on program states. We demonstrate that code reuse attacks violate these requirements and thus can be detected; further, new exploit variations will not circumvent $\mathcal{CPE}$. To provide an efficient and adoptable solution, we also define *observed conformant program execution*, which validates program state at system call invocations; we demonstrate that this relaxed model is sufficient to detect code reuse attacks. We implemented our algorithm in a tool, ROPStop, which operates on unmodified binaries, including running programs. In our testing, ROPStop accurately detected real exploits while imposing low overhead on a set of modern applications: 5.3% on SPEC CPU2006 and 6.3% on an Apache HTTP Server.

Keywords: Binary analysis, static analysis, return-oriented programming, jump-oriented programming.

1   Introduction

Code reuse attacks are an increasingly popular technique for circumventing traditional program protection mechanisms such as $W \oplus X$ (e.g., Data Execution Prevention (DEP)), and the security community has proposed a wide range of approaches to protect against these attacks. However, many of these approaches provide ad hoc solutions, relying on observed attack characteristics that are not intrinsic to the class of attacks. In the continuing arms race against code reuse attacks, we must construct defenses using a more systematic approach: good engineering practices must combine with the best security techniques.

Any such approach must be engineered to cover the complete spectrum of attack surfaces. While more general defensive techniques, such as Control Flow

J. Jürjens, F. Piessens, and N. Bielova (Eds.): ESSoS 2014, LNCS 8364, pp. 1–18, 2014.

Integrity or host-based intrusion detection, provide good technical solutions, each is lacking in one or more features necessary to provide a comprehensive and adoptable solution [51]. We must develop defenses that can be effectively applied to real programs.

We present a technique based on first principles for the efficient, robust detection of code reuse attacks. Our work is grounded in a model of conformant program execution ($\mathcal{CPE}$), in which we define what program states are possible during normal execution. We generate our model automatically from the program binary; thus, no learning phase or expert knowledge is required. $\mathcal{CPE}$ enforces expected program behavior instead of defending against anticipated attacks; thus, new exploit variations will not circumvent $\mathcal{CPE}$. $\mathcal{CPE}$ is based on observable properties of the program counter and runtime callstack; a program has $\mathcal{CPE}$ if, for all program states during the execution of the program, the program counter and callstack are individually valid and consistent with each other. Code reuse attacks execute short sequences of instructions without respect to their location in original code; thus, these attacks deviate from our model.

Conformant program execution verifies each program state; therefore, continually validating it can result in high overhead. We address this problem with *observed conformant program execution* ($\mathcal{OCPE}$), which reduces overhead by only validating program state at system call executions. We demonstrate that this relaxed model is sufficient to detect code reuse attacks. Thus, $\mathcal{OCPE}$ provides an adoptable solution while still providing safety guarantees. $\mathcal{OCPE}$ is not designed to handle code reuse-based mimicry attacks, which have not yet been demonstrated in the research world or seen in the wild. We believe $\mathcal{OCPE}$ could be augmented in future work to handle these attacks.

We engineer our approach using a component model based on strong binary analysis of the code. This analysis allows us to operate on modern binaries, which frequently are highly optimized or lack debugging information; our analysis does not rely on information that may not be present in a modern application. We leverage a binary analysis toolkit to identify key characteristics of the stack frame at any instruction in the binary; this allows us to gather a full callstack via a *stackwalk* at runtime. While conceptually straightforward, accurate stackwalks are surprisingly difficult to perform on real applications. Our algorithm leverages these stackwalks, taken at system calls, to reliably detect code reuse attacks.

We implemented our code reuse detection algorithm in a tool, ROPStop, which operates on unmodified binaries, including running programs. We evaluated ROPStop using real exploits from two classes of code reuse attacks: return-oriented programming (ROP) and jump-oriented programming (JOP). Our results show that our tool is able to correctly identify each exploit. We tested ROPStop with the SPEC CPU2006 benchmarks and an Apache HTTP Server as a control group of unexploited, conventional binaries to evaluate overhead and measure the occurrence of false positives. Our results show an average overhead of 5.3% on and 6.3% on Apache; ROPStop reported no false positives.

We provide an overview of the challenges in detecting code reuse attacks and existing work in this area in Section 2 and a formal description of conformant program execution and code reuse attacks in Section 3. Next, we describe the technical details of our approach in Section 4. We evaluate our approach in Section 5 and finish with a brief conclusion in Section 6.

2 Background and Related Work

Code reuse attacks provide interesting new challenges for security researchers. While $W \oplus X$ guards against code injection attacks, it is insufficient to stop code reuse attacks because such attacks do not write new code into the address space. We describe how an attacker gains control of the program and produces a code reuse attack. Further, we describe how an attacker locates gadgets within the program. We conclude with a discussion of techniques that are not specifically focused on code reuse attacks but are similar to techniques presented here.

2.1 Gaining Control of the Program

The first step of a code reuse attack is to gain control of the program counter to divert program control flow to the first gadget. This is done by making use of an existing vulnerability (e.g., a buffer overflow) to alter program data. Although these vulnerabilities and possible defenses are well studied, attackers remain able to exploit these vulnerabilities and launch attacks [51]. We assume that an attacker will be able to find a viable entry point for launching a code reuse attack, and do not discuss these vulnerabilities further. To ensure that program control flow will be diverted, an attacker overwrites either the return address for the calling function or a function pointer with the address of the first gadget. Note that $W \oplus X$ restricts attackers to cases where control flow targets depend on writeable locations rather than executable locations.

If the return address is overwritten, control flow will be diverted to the gadget when the current function returns and the new return address is loaded into the program counter. Stack-smashing attacks such as this one have been well-studied and techniques such as StackGuard [16] will prevent these attacks. To be effective, this protection must be present in *all program code*, including the program binary and any libraries on which it depends. These protections are provided as options by most modern compilers, but they are frequently not turned on by default, and can be turned off. Therefore, it is not safe to assume these protections will be present, and frequently it is not possible for a user to modify shared libraries (e.g., `libc.so`) to include them. If a function pointer is overwritten, control flow will be diverted to the gadget when the program invokes an indirect call or jump using the address of the function pointer.

Once control flow has been diverted to the first gadget, the code reuse attack begins. We note that an attack might also use an existing vulnerability to modify program data: for instance, to cause a system call to be executed with unintended arguments. These data driven attacks [2, 14, 20] are complementary to control flow-based attacks and are beyond the scope of our work.

2.2 Gadget Execution

Return- and Jump-Oriented Programming. In ROP, each gadget is terminated with a return instruction [44]. Thus, if an attacker has gained control of the stack pointer, they can use these return instructions to cause program execution to flow from one gadget to the next. JOP attacks use the same general technique, but gadgets are chained together with indirect jump instructions [7,9]. Thus, unlike ROP, JOP does not rely on manipulating the stack pointer; instead, indirect jump instructions have a specified target location, often stored in a register. This provides an extra challenge in constructing a jump-oriented attack: gadgets must manipulate relevant register values to ensure each indirect jump transfers control to the next gadget.

Defenses against Code Reuse Attacks. There are a variety of techniques designed to detect code reuse attacks. Several approaches make use of a shadow stack to prevent control flow manipulation that relies on overwritten stack values [12, 19, 23]. Others try to detect gadget execution by monitoring the length of instruction sequences between returns [11, 18]. Still others proposed monitoring pairs of call and return instructions [12, 31]. These approaches each target return-oriented attacks; however, they will not detect jump-oriented attacks because these attacks do not rely on return instructions to transfer control between gadgets. Another common approach ensures that a function should only start executing at its entry point [12, 29–31]. However, an attack may still hijack control flow while conforming to these requirements, thus remaining undetected.

Each of these mitigation techniques targets specific characteristics of previously observed code reuse attacks; such features may not be intrinsic to all code reuse attacks. In contrast, our work detects any violations of our model of conformant program execution by validating known properties of the program. Thus, evolving attack variations will not hinder our ability to detect these attacks.

Still other techniques monitor system calls to detect violations. ROPGuard [24] is a recent tool focused on ROP attacks that checks for a valid callstack at a subset of system calls. ROPGuard relies on frame pointers to traverse the callstack. Many modern programs do not save frame pointers; thus, callstack verification is turned off by default, leaving such programs vulnerable. By verifying conformant program behavior at *all* system calls and using a robust binary analysis toolkit to perform accurate stackwalks, our work provides a more complete approach.

2.3 Gadget Discovery

Locating potential gadgets is performed by scanning a target binary for return instructions (for ROP) or indirect jumps (for JOP). An attacker then chooses the gadgets to use from this set potential gadgets [13, 21, 47]. This selection of gadgets must allow the attacker to maintain command over the control flow of the program and perform desired actions while avoiding unwanted side effects.

ASLR is a common system-level technique that randomizes the addresses at which libraries are loaded into the address space of a process; this is particularly relevant for code reuse attacks because these attacks rely on known locations for each gadget. Unfortunately, this is not sufficient to prevent a code reuse attack. Schwartz et al. point out that not all operating systems randomize all components within the address space, and some require an application to explicitly turn on ASLR [47]. As long as there exists some segment of code that is not randomized, code reuse attacks are possible. Checkoway et al. demonstrated that such attacks can be constructed even with a limited set of instructions [10]. Consequently, in real systems, we must assume that if gadgets exist in the code, an attacker will be able to find a sufficiently powerful set to perform their attack.

Several prevention techniques attempt to eliminate possible gadgets in library code via code diversification [17,28,32,36,53]. An alternative prevention strategy seeks to create binaries or kernels that lack necessary characteristics for ROP attacks [34,35]. Many software diversification techniques rely on modifying the program, library, or kernel binaries via recompilation or binary rewriting; such modifications may not be possible in real systems. Furthermore, these techniques do not preclude code reuse attacks, but simply challenge attackers to identify gadgets in more sophisticated ways [51]. In contrast, ROPStop is engineered to provide a comprehensive solution that does not require ASLR or recompilation; ROPStop operates on unmodified binaries, including running programs.

2.4   Other Approaches

Our work is also similar to techniques that, while not focused on code reuse attacks, may be effective against such attacks: control flow validity enforcement and anomalous system call detection. Control-Flow Integrity (CFI) ensures that program execution holds to a control-flow graph (CFG) derived from static analysis [1]. However, because CFI verifies each control-flow transfer during program execution, it imposes high runtime overhead. More practical approaches, such as Control Flow Locking (CFL) [6], Compact Control Flow Integrity and Randomization (CCFIR) [54], and CFI for COTS binaries [55], have other limitations. CFL lazily verifies transfers, which greatly improves performance; however, their technique requires statically-linked binaries, which severely limits its application. CCFIR and CFI for COTS use binary rewriting to add verification checks at indirect control flow transfers. Although these approaches can be applied to shared libraries, protections must be applied to *all* binary code to ensure the implied security guarantees; any unprotected code is a potential attack target. Thus, the user applying protections must both be aware of and able to protect all library dependencies; this requirement can limit the applicability of these approaches. Further, CCFIR relies on ASLR for all program code. The limitations of these techniques prevent them from providing comprehensive defense solutions.

Host-based intrusion detection systems (IDS) use anomalous patterns of system calls to identify attacks. These approaches rely on a learning phase [22,25,48] or static binary analysis [26]. Unlike learning-based IDS, our work is based on a model of what program states are possible in normal execution. Further, our approach enforces a valid program state at each system call, rather than a valid pattern of system calls. We note that mimicry attacks [33,42] allow an attacker to subvert system call monitoring by ensuring that both the call stack of each system call and the sequence of system calls made by a compromised program appear normal to an IDS. However, mimicry attacks rely on code injection. While it is theoretically possible to extend a mimicry attack to employ code reuse, this form of attack does not appear in the wild; this would require an attacker to construct a sequence of gadgets that both executes their attack and restores the system to a state that appears valid at the next system call. However, gadget discovery is a difficult problem. In practice, attackers who employ code reuse are attempting to find the shortest route to a less restrictive environment [43]. Thus, we consider mimicry attacks beyond the scope of our work.

3   Conformant Program Execution

Our work is grounded in a model of conformant program execution. We create a definition of program state and then define requirements on that program state that must hold true at runtime. Program executions for which these requirements hold true are called *conformant* executions. Any deviation from these requirements during program execution indicates a non-conformant execution. We first describe our notation, then define a model of conformant program execution, and finally discuss how code reuse attacks will be detected as non-conformant.

3.1   Notation

A program is a set of procedures, $P = \{p_1, \ldots, p_m\}$. I is the set of all possible machine instructions; $I_P \subseteq I$ are the valid instructions for P. Each procedure p_j is a tuple, $p_j = (I_{p_j}, entry_{p_j}, \{exit_{p_j}\})$, where $entry_{p_j}$ and $exit_{p_j}$ are instructions in I_{p_j} that represent the entry point and the zero or more exit points for p_j. To represent valid interprocedural control flow in P, we define a call multigraph, $CMG_P = (N_P, E_P)$, where $N_P = P$ and each $e = (p_s, p_t, i) \in E_P$ is a control flow transfer $p_s \rightarrow p_t$ at i where $i \in I_{p_s}$ is the instruction that effects the call.

We parse the program binary using control and dataflow analysis; these analyses produce a call graph and control flow graph, which are used to populate the data structures in our model. More details are presented in Section 4. The resulting CMG_P may be incomplete due to unknown indirect control flow at $i \in p_j$. We address this incompleteness by making conservative assumptions about this control flow, such as that any procedure may be the target of an indirect call. In such cases, additional edges $(p_j, *, i)$ may be added to CMG_P, where $i \in p_j$ and the target of the control flow transfer at i is unknown. For increased accuracy, CMG_P could be augmented at runtime using dynamic analysis [46]; however, there is an increase in overhead associated with such analysis.

We define an execution of P as $execution(P) = \langle m_1, \ldots, m_n \rangle$, where each m_i represents an instance of program state. Program state includes elements of machine state that are affected by program execution, including registers and memory. We represent two elements of interest, the program counter and the callstack, with $pc(m)$ and $callstack(m)$. The callstack, C, is a sequence of currently active stack frames, $C = \langle s_0, \ldots, s_n \rangle$, where $s_{top} = s_n$ is the frame at the top of the stack associated with the currently active procedure.

We refer to a stack frame as a tuple $s = (p, i, height)$, where p refers to the procedure associated with the stack frame. i refers to the last executed instruction in the context represented by s. For the top stack frame, this instruction is the program counter; for all other frames, this instruction is the call immediately before the frame's return address (Figure 1). $height$ is the current height of the stack frame. Stack frame height is based on the space needed to store saved registers, procedure parameters, local variables, and a return address. Because different parts of a procedure may require different amounts of local storage, this height may vary for different parts of the procedure. Therefore the control flow path leading to the current location within a procedure, which we denote $entry_{proc(s)} \rightsquigarrow instr(s)$, affects the stack frame height. Our model does not rely on other information, such as a saved frame pointer, because such information may be omitted by optimizing compilers. We represent elements of each frame with $proc(s)$, $instr(s)$, and $height(s)$.

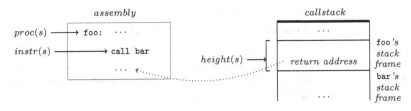

Fig. 1. Relationship between the tuple $s = (p, i, height)$ that represents the current stack frame instance for procedure `foo`, the assembly code for `foo`, and the callstack. Elements of the stack frame are represented with $proc(s)$, $instr(s)$, and $height(s)$. Note that the choice to associate the return address with the caller frame and not the callee frame is arbitrary; this could be restated without loss of generality.

We further define $Heights(i)$ to represent the set of *valid heights* possible from all paths through the CFG from $entry_{p_j}$ to $i \in p_j$. If code modifies stack frame height inside a loop, the height will depend on the number of loop iterations, so it is possible that the size of the set is not finite. In such cases, we assume that the stack frame height at these instructions is unknown. In principle, we could analyze the instruction sequence and build a closed form model of such behavior. In practice, this occurrence is exceedingly rare.

3.2   Conformant Program Execution

Our model of conformant program execution should be permissive enough to allow a program to execute valid program instructions reached via valid control flow, while restrictive enough to allow only these valid executions and detect code reuse attacks. We define a program to have conformant execution based on characteristics of two program state components: the program counter pc

and the runtime callstack C. A program is conformant at a given time, i.e., a particular machine state m_i, if each component is individually valid and if both components are consistent. A program has conformant program execution ($\mathcal{CPE}$) if the program is conformant for all program states in $execution(P)$.

To verify conformance for a particular program state during execution, we examine the program counter pc and the callstack C. The program counter should contain the address of a valid instruction. An instruction j is valid if it exists in the set of valid instructions for program P; $valid(i) : i \in I_P$. The program counter is valid if it points to a valid instruction; $valid(pc) : valid(instr(pc))$. This requirement eliminates the use of unaligned instructions that could provide a rich selection of unintended instruction sequences to be used in an attack [8]. Further, should $W \oplus X$ not be in place, this requirement precludes code injection attacks, which rely on code located outside of valid code sections of the binary.

A callstack $C = \langle s_0, \dots, s_n \rangle$ is valid if a height requirement holds for each frame in C and if a call requirement holds for each pair of adjacent frames.

$$valid(C) : \forall\ s \in C : valid(s)$$

$$valid(s_k) : \begin{cases} valid\_height(s_k) & k = n \\ valid\_height(s_k) \wedge valid\_call(s_k) & 0 \le k < n \end{cases}$$

A stack frame has a valid height h if that height is a member of the set of valid heights $Heights(i)$ for the corresponding instruction. For each pair of adjacent stack frames to meet the call requirement, the control transfer represented by each pair must correspond to an edge in the call multigraph.

$$valid\_height(s_k) : height(s_k) \in Heights(instr(s_k))$$

$$valid\_call(s_k) : (proc(s_k), proc(s_{k+1}), instr(s_k)) \in E_P$$

Validating calls between procedures associated with consecutive stack frames ensures that C represents a valid interprocedural control flow path through P. Thus, we incorporate the goals of CFI [1] but perform verification with runtime checks for an efficient implementation. Fratric also proposed a requirement on callstack validity [24]. In his work, a valid callstack requires each stack frame to have a valid return address and a valid call target, though the tool, ROPGuard, is only able to implement the former because it relies on information often not present in modern binaries. We discuss these technical differences in Section 4.

Finally, we define what is required for the program counter and the callstack to be consistent. Given a program counter that points to a valid instruction and a valid callstack, we need to ensure they are mutually consistent; that is, that the callstack is valid for that program counter: $consistent(i, C) : i = instr(top(C))$. These validity checks determine if a program has conformant execution:

$$valid(m) : valid(pc(m)) \wedge valid(callstack(m)) \wedge consistent(pc(m), callstack(m))$$

$$\mathcal{CPE}(P) : \forall\ m \in execution(P) : valid(m)$$

3.3   Code Reuse Attacks

We introduce a notation for code reuse attacks and discuss how our model detects these attacks. A code reuse attack is comprised of a series of gadgets,

$A = \langle g_0, \ldots, g_n \rangle$. Each gadget is a tuple: $g_k = (I_{g_k}, entry_{g_k}, exit_{g_k})$, where $entry_{g_k}$ and $exit_{g_k}$ are the instructions in I_{g_k} that represent the entry point and exit point for g_k. The instructions that comprise each gadget are part of the set of possible machine instructions I, but not necessarily part of I_P [49].

Although we provide a broad definition of gadgets that spans several flavors of code reuse attacks, there are several additional known characteristics of each gadget. Such characteristics are often used when crafting code reuse attacks [13, 47]. The last instruction in a gadget, $exit_{g_k}$, must cause a control flow transfer to the next gadget g_{k+1}. Gadgets g_k and g_{k+1} are chosen such that $exit_{g_k}$ transfers control to $entry_{g_{k+1}}$. In ROP, each $exit_{g_k}$ is a return instruction and the address of $entry_{g_{k+1}}$ is located at the top of the stack [49]; in JOP, $exit_{g_k}$ is an indirect jump instruction, where the target of the jump is $entry_{g_{k+1}}$ [7, 9].

Executing a sequence of gadgets interrupts the original execution of the program. However, gadget execution occurs in the context of the program and may use the original program stack. As a result, gadgets typically violate $\mathcal{CPE}$ in one or more of three ways: executing invalid instructions, altering the stack, or not following the original control flow of the program. To detect gadget execution, and thus a code reuse attack, we must identify these violations.

First, gadgets may execute invalid instructions, either code that was injected or misaligned instructions from the original program. We detect these invalid instructions by examining the program counter: $\forall\, g_k \in A : I_{g_k} \subseteq I_P$.

Second, gadgets may include instructions that alter the stack. Let $C = \langle s_0, \ldots, s_m \rangle$ represent the callstack prior to the execution of the first gadget, where s_m represents the top of the stack. While a gadget g_k executes, $C_k = \langle s_0, \ldots, s_n, \ldots, s_q \rangle$ where $\langle s_0, \ldots, s_{n-1} \rangle$ represents what remains of the original callstack and $\langle s_n, \ldots, s_q \rangle$ represents the effects of executing gadget code.

When g_0 begins execution, $s_n = s_m$. If subsequent gadgets remove values from the stack, $\langle s_0, \ldots, s_{n-1} \rangle \sqsubseteq \langle s_0, \ldots, s_m \rangle$, where $A \sqsubseteq B$ means that A is a contiguous subsequence of B that either starts at the same initial state (i.e., s_0) or is empty. Otherwise, $\langle s_0, \ldots, s_{n-1} \rangle = \langle s_0, \ldots, s_{m-1} \rangle$.

Each gadget must maintain the correct height in the context of the stack frame in which it executes; the expected height for the first instruction in each gadget, $entry_{g_k} \in I_{p_j}$, must match the height of s_n. Otherwise, the height of frame at the top of the stack will be observably incorrect at all instructions in the gadget. This invariant must be maintained as subsequent gadgets execute. We detect these invalid stack frames by verifying the height of each stack frame:

$$\forall\, i \in I_{g_k}, g_k \in A : \; height(entry_{g_k} \leadsto i) + height(s_n) = height(entry_{p_j} \leadsto i),$$

$$\text{where } I_{g_k} \in I_{p_j}$$

Third, gadgets may induce a control flow path that does not match that of the original program. The instructions that comprise each gadget are selected from an existing procedure; to maintain conformant execution, there must be a valid control flow transfer from the procedure corresponding to the stack frame below s_n to the procedure that contains the gadget's instructions. We detect these invalid control flow paths by validating the path represented by the callstack:

$$\forall\, g_k \in A : (proc(s_{n-1}), p_j, instr(s_{n-1})) \in E_P, \text{ where } I_{g_k} \in p_j$$

However, there are two ways in which an attack might be able to conform to our model. First, there may be cases in which a loop allows multiple valid stack heights at a particular instruction, as described in the previous section; in this case, a gadget may be able to use an invalid stack height without detection, i.e.,

$$|Heights(instr(s_k))| > 1 \ \wedge \ height(s_k) \in Heights(instr(s_k)) \ \wedge$$

$$height(s_k) \neq height(entry_{proc(s_k)} \rightsquigarrow instr(s_k))$$

In our experience, such code is rare, and can be mitigated by unrolling or otherwise modifying the loop such that $|Heights(instr(s_k))| = 1$. Second, an attack might take advantage of our conservative handling of indirect control flow to construct unintended paths through the binary via indirect procedure calls, i.e.,

$$(proc(s_{k-1}), *, instr(s_{k-1})) \in E_P \ \wedge \ proc(s_{k-1}) \not\rightarrow proc(s_k) \text{ at } instr(s_{k-1})$$

Such an attack would have to make use of existing program code using whole procedures at a time, because executing a sequence of procedure fragments would result in one or more invalid stack frames. Such attacks must still override the targets of one or more indirect calls to force the execution of the program down the path desired by the attack. For example, an attack could begin by overwriting a table of function pointers. However, this is made more complex by the fact that it must not only overwrite the function pointer but also overwrite the data used to set up the parameters used at the callsite. While this strategy seems unlikely, as we use more sophisticated analysis to refine indirect call edges, we further reduce the likelihood of circumventing our technique.

3.4   Observed Conformant Program Execution

In this section, we define observed conformant program execution ($\mathcal{OCPE}$), which improves efficiency by verifying program states only at system call entries. Monitoring at system call granularity offers two advantages. First, the system call tracing interface provided by operating systems (e.g., PTRACE_SYSCALL) cannot be interrupted; therefore, we will receive notification of system call entry even if an attack is in progress. Second, attacks must use the system call interface (e.g., exec) to modify the overall machine state; therefore, we will not miss the effects of an attack by monitoring only at system calls. We define $\mathcal{OCPE}$ as follows:

$$is\_syscall(m) : callstack(m) = \langle s_0, \ldots, s_n \rangle; proc(s_n) \text{ invoked system call } syscall(m)$$

$$observed(P) = \langle m \in execution(P) \ | \ is\_syscall(m) \rangle$$

$$\mathcal{OCPE}(P) : \ \forall \ m \in observed(P) : \ valid(m)$$

$\mathcal{OCPE}$ makes the assumption that the effects of an attack will be visible in the program state at the point a system call is executed. As discussed in Section 2, current code reuse attacks are not constructed to evade $\mathcal{OCPE}$ [13,47]. We demonstrate in Section 5 that $\mathcal{OCPE}$ is effective against these attacks. Furthermore, we believe it will be difficult for future code reuse attacks to entirely hide their effects from our stack model, because an attack would have to both hide its own effects as well as forge a consistent program state. Even if a future attack could circumvent $\mathcal{OCPE}$ in this way, our model of $\mathcal{CPE}$ would detect these deviations from normal program execution. Therefore, $\mathcal{OCPE}$ is an effective optimization of $\mathcal{CPE}$ that greatly improves performance while preserving the power of $\mathcal{CPE}$.

4   Implementation

We have incorporated our model into a tool, ROPStop, that monitors a process during its execution for observed conformant program execution. We use several components from the Dyninst binary modification and analysis toolkit to perform this runtime monitoring and verification [37], and ROPStop has been implemented for both 32- and 64-bit x86/Linux. In total, ROPStop is approximately 3,000 lines of C++ on top of several toolkit libraries.

4.1   Process Monitoring

We perform runtime process monitoring using the ProcControlAPI process monitoring and control component [40] of Dyninst. This library allows a tool process (ROPStop) to manage one or more target processes using platform-independent abstractions. ProcControlAPI includes the ability to control threads, set breakpoints, request notifications (callbacks) at events, and read and write memory.

Using ProcControlAPI, we may either create a new process or attach to a running process. We then register a callback function that is invoked before each system call. Although ProcControlAPI already provides the capability to register callback functions at various process events using the operating system's debug interface (ptrace), we extended the library to provide support for callbacks at system call entry. ROPGuard [24] used an alternative approach that instead operates on the library wrappers for system calls rather than directly on the system call. However, this can be defeated by malicious code that directly executes a trap instruction to invoke a system calls By using the debug interface, we can guarantee that all system calls will trigger our callback.

ROPStop then parses the program binary and any library dependencies using the ParseAPI control flow analysis library [39]. ParseAPI uses recursive traversal parsing to construct a whole-program control flow graph [15,52], including sophisticated heuristics to recognize functions that are only reached by indirect control flow and works in the absence of symbol table information [27,45]. This CFG provides both the call multigraph (CMG_P) required by our model and the information necessary to identify each valid program instruction (I_P).

Once the process binary has been parsed, we continue its execution. If the process creates additional threads or launches a new process, ProcControlAPI will monitor these also. A monitored process is stopped at each system call entry. ROPStop checks that the program has conformant execution; ROPStop explicitly verifies the program counter and the callstack and implicitly verifies consistency. If the process is conformant, execution is continued. If ROPStop detects non-conformant program state, the process is terminated.

4.2   Instruction Validity

ROPStop first verifies that the program counter points to a valid instruction in the original program, e.g., that $instr(pc) \in I_P$. This inexpensive step ensures that an attack may not make use of unaligned instructions. We identify the basic block in the CFG that contains this address and disassemble the block using the InstructionAPI instruction disassembly library [38]. If we reach an instruction that begins at the address in the program counter, we conclude the current instruction is valid. Otherwise, we conclude the instruction is invalid.

4.3 Callstack Validity

Next, ROPStop gathers a full stackwalk using the StackwalkerAPI library [41]. A full stackwalk must be gathered at each system call to ensure that no malicious stack modifications have occurred since the last validity check; this step must be completed even if two system calls occur within the same function or a single loop. ROPStop uses StackwalkerAPI to walk the stack one frame at a time and validate each frame before continuing the stackwalk. StackwalkerAPI represents stack frames as pairs (r, sp), where r is a location in the program and sp is the value of the stack pointer in that location. Given an input frame (r_i, sp_i), StackwalkerAPI calculates the expected previous frame (r_{i-1}, s_{i-1}).

Given a valid stack frame s_j and an expected previous frame s_{j-1} from StackwalkerAPI, we validate it as follows. We verify the height of the stack frame s_j by validating the return address of s_{j-1}, r_{j-1}. We define a return address to be valid if the instruction at this address is immediately preceded by a call instruction; this definition is also used by the ROPGuard tool [24]. ROPStop uses InstructionAPI to locate the instruction prior to the return address and identify its type. If s_{j-1}'s return address is valid, we conclude that the height of s_j is valid, e.g., that $height(s_j) \in Heights(instr(s_j))$.

This validation allows ROPStop to verify the second necessary condition for a valid stack frame: that there exists a valid control flow transfer between the caller (s_{j-1}) and the current, callee frame (s_j). We check the CFG constructed by ParseAPI to verify that there is a valid call edge from the caller to the callee; this edge must originate at the call instruction found in the previous step. This verifies that $(proc(s_{j-1}), proc(s_j), instr(s_{j-1})) \in E_P$.

Performing a stackwalk is not always an easy task. The debugging information that describes stack frames (e.g., DWARF call frame information) is frequently missing; for example, commercial binaries frequently omit this information due to concerns about reverse engineering. Even if this debug information is present it is frequently incomplete or incorrect; for example, compilers often omit stack-walking information for automatically generated code.

In light of these challenges, we extended StackwalkerAPI to use dataflow analysis to identify stack heights. This analysis begins at the entry to a function and tracks the effects of all instructions that modify the stack heights; the result is the set of possible stack heights for each instruction in the function. This robust analysis enables an accurate stackwalk in the absence of debugging information.

If StackwalkerAPI reaches the bottom of the stack and does not encounter an invalid stack frame, then we conclude that the callstack is valid. Otherwise, we have found non-conformant program state, and the process is terminated.

5   Evaluation

ROPStop provides protection against code reuse attacks while identifying no false positives. We verified these characteristics with the following experiments. First, we tested ROPStop against code reuse attacks, as well as a conventional stack smashing attack, and show that ROPStop detects each of these attacks. Second, we tested ROPStop against a set of conventional binaries, and show that ROPStop results in no false positives while imposing overhead of only 5.3% on SPEC benchmarks and 6.3% on an Apache HTTP Server.

All evaluation was conducted on a 2.27GHz quad-core Intel Xeon with 6GB RAM, running RHEL Server 6.3 (kernel 2.6.32). All exploits were run inside VirtualBox 4.2.0 virtual machines, running Debian 5.0.10 (2.6.32) or Ubuntu 10.04 (2.6.26); see Table 1. SPEC and Apache were run directly on the host.

Table 1. Details about each exploit and results for ROPStop's detection of the attack. All attacks were detected because of invalid stack frame heights; the exploit characteristic that lead to the observed invalid program state is also provided.

Name	OS	Exploit Source	Detected	Why Invalid
17286a (ROP)	Ubuntu 10.04	sickness [50]	✓	Overwritten return address
17286b (ROP)	Ubuntu 10.04	sickness [50]	✓	Overwritten return address
Rsync (ROP)	Debian 5.0.10	Schwartz et al. [47]	✓	Overwritten return address
Bletsch (JOP)	Debian 5.0.10	Bletsch et al. [7]	✓	Gadget executing
Stack-smash	Debian 5.0.10	Aleph One [3]	✓	Overwritten return address

We began by testing ROPStop against known attacks; we summarize each attack in Table 1. The 17286 ROP exploits are from http://www.exploit-db.com. Rsync is a ROP exploit generated by the Q tool [47]. We acquired the code necessary for this exploit from the authors; additional exploits from this work were unavailable. Bletsch is a JOP attack [7]. Finally, we include a canonical buffer-overflow attack, Stack-smash, to demonstrate ROPStop's ability to detect these attacks also. We used information provided in the original documentation to create the vulnerable program as well as the input necessary for each exploit.

The results of testing ROPStop against these attacks are shown in Table 1. We were successful in detecting each attack. In each case, callstack verification failed; the height of a stack frame was found to be invalid because the return address in the expected caller frame did not follow a `call` instruction.

Next, we used the SPEC CPU2006 benchmark suite and an Apache HTTP Server (version 2.4.6) [5] to provide a control group of conventional binaries. We applied ROPStop to the execution of these binaries to both detect any false positives and determine how much overhead ROPStop incurs. We selected SPEC because the execution of each benchmark is well understood and as CPU intensive programs, any overhead imposed by ROPStop would not be hidden by I/O. Each benchmark was run three times with the reference inputs; we report the mean of these runs. We measured end-to-end times, which includes the time required to generate our model as well as the runtime of the program.

ROPStop generated no false positives when run on the SPEC CPU2006 benchmark suite. The overhead imposed by ROPStop is shown in Figure 2; on average,

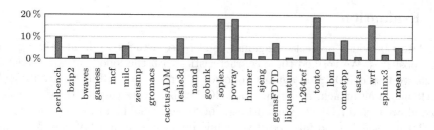

Fig. 2. Overhead results for SPEC CPU2006 benchmarks under ROPStop; **mean** represents the geometric mean of all overhead values. We omit four benchmarks, `gcc`, `calculix`, `dealII`, `xalancbmk`; we were unable to successfully run these unmonitored.

ROPStop imposes 5.3% overhead. In most cases ROPStop imposes under 3% overhead; our highest overhead is 19.1% for `tonto`.

Table 2. Full results for SPEC CPU2006 benchmarks under ROPStop. The system call rate is reported as system calls per second based on unmonitored runtimes; components of overhead are reported as percentages of total overhead imposed (summing to 100%).

Benchmark	System Call Rate (calls/second)	% Overhead Imposed	Overhead Breakdown		
			% Instruction Validity	% Callstack Validity	% Context Switching
perlbench	167.8	9.6	0.2	50.7	49.1
bzip2	2.0	0.8	0.0	43.8	56.2
bwaves	3.3	1.4	0.0	31.7	68.2
gamess	29.5	2.4	0.1	59.0	40.9
mcf	3.4	1.9	0.0	51.4	48.5
milc	25.5	5.7	0.1	23.5	76.4
zeusmp	0.2	0.7	0.0	54.2	45.7
gromacs	1.5	0.6	0.0	45.1	54.9
cactusADM	7.6	1.0	0.1	52.0	47.9
leslie3d	31.4	9.2	0.1	14.2	85.7
namd	3.1	0.9	0.0	61.6	38.4
gobmk	14.0	2.2	0.1	43.6	56.4
soplex	241.5	18.1	0.2	50.8	49.1
povray	156.2	18.1	0.1	53.3	46.6
hmmer	18.6	2.6	0.1	38.8	61.1
sjeng	4.8	1.3	0.0	40.5	59.5
GemsFDTD	88.7	7.3	0.2	40.9	59.0
libquantum	0.3	0.7	0.0	50.2	49.8
h264ref	5.1	1.3	0.0	28.9	71.0
tonto	119.6	19.1	0.1	41.2	58.7
lbm	1.4	3.4	0.0	15.5	84.5
omnetpp	3.7	8.8	0.0	12.7	87.3
astar	7.2	1.3	0.1	56.7	43.3
wrf	53.2	15.7	0.0	13.4	86.6
sphinx3	18.6	2.4	0.1	38.1	61.8

The overhead imposed by ROPStop is dependent on the frequency of system calls as well as the height of the stack. ROPStop is a separate process, rather than in the same address space as the monitored application. Validating state at each system call requires at least two context switches; these context switches are more expensive than the validation checks ROPStop performs. Thus, benchmarks that make frequent use of system calls or, to a lesser extent, have deep call stacks, suffer higher overhead. We report the breakdown of this overhead in Table 2. The SPEC benchmarks vary greatly in the number of invoked system calls. The average number of system calls is 23,635; `zeusmp` invokes only 140, while `tonto` invokes 153,890 system calls. We omit four benchmarks, `gcc`, `calculix`, `dealII`, and `xalancbmk`, because we were unable to run them unmonitored; we expect these to have similar performance to our reported numbers.

We used the ApacheBench tool [4] to measure the performance of the Apache server while being protected by ROPStop. We ran ApacheBench with various total numbers of requests to a local, static page; the number of concurrent requests was always 100. Each request size was run twenty times; we report the mean of these runs. For each run, we started the server, attached ROPStop, and then ran ApacheBench. We report numbers recorded by ApacheBench, rather than end-to-end results, because web servers are commonly long-running applications; however, model generation times were similar to those for SPEC. ROPStop generated no false positives when run on the Apache server. The overhead imposed is shown in Figure 3; on average, ROPStop imposes 6.3% overhead. The Apache server made between approximately 3,000 and 60,000 system calls for the smallest and largest total number of requests, respectively.

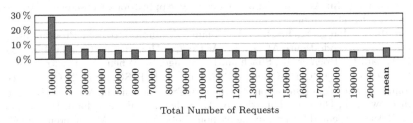

Total Number of Requests

Fig. 3. Overhead results for an Apache HTTPD web server run under ROPStop, measured using ApacheBench; **mean** represents the geometric mean of all overhead values.

6   Conclusion

We have presented a systematic approach for detecting code reuse attacks. In contrast to current techniques, which rely on exploit characteristics that may not be intrinsic to the class of attacks, our approach is based on first principles. We defined a model of conformant program execution; by verifying the program counter and callstack, we were able to detect code reuse attacks, even as they continued to evolve. To provide an efficient and adoptable solution, we also defined observed conformant program execution, which provided the same guarantees while only verifying program conformance at system calls. Finally, we built a tool, ROPStop, that uses our model of observed conformant program execution to detect code reuse attacks. Our results show that ROPStop is capable of accurately detecting real code reuse attacks. Further, when tested on a set of modern applications, ROPStop produced no false positives and incurred only 5.3% overhead on SPEC CPU2006 and 6.3% on an Apache HTTP Server.

Acknowledgments. This work is supported in part by Department of Energy grants DE-SC0003922 and DE-SC0002154; National Science Foundation Cyber Infrastructure grants OCI-1032341, OCI-1032732, and OCI-1127210; and Department of Homeland Security under AFRL Contract FA8750-12-2-0289.

References

1. Abadi, M., Budiu, M., Erlingsson, U., Ligatti, J.: Control-flow integrity principles, implementations, and applications. ACM Trans. Info. & Systems Security (TISSEC) 13, 4:1–4:40 (2009)
2. Akritidis, P., Cadar, C., Raiciu, C., Costa, M., Castro, M.: Preventing Memory Error Exploits with WIT. In: IEEE Symposium on Security and Privacy, Oakland, CA (May 2008)
3. Aleph One: Smashing the stack for fun and profit. Phrack Magazine 7(49), 14–16 (1996)
4. Apache Software Foundation: ab - Apache HTTP server benchmarking tool (July 2013), http://httpd.apache.org/docs/2.2/programs/ab.html
5. Apache Software Foundation: Apache HTTP Server Project (July 2013), http://www.apache.org
6. Bletsch, T., Jiang, X., Freeh, V.: Mitigating Code-Reuse Attacks with Control-flow Locking. In: Annual Computer Security Applications Conference (ACSAC), Orlando, FL (December 2011)
7. Bletsch, T., Jiang, X., Freeh, V.W., Liang, Z.: Jump-Oriented Programming: A New Class of Code-Reuse Attack. In: ASIACCS, Hong Kong, China (March 2011)
8. Buchanan, E., Roemer, R., Shacham, H., Savage, S.: When Good Instructions Go Bad: Generalizing Return-Oriented Programming to RISC. In: ACM Conference on Computer and Communications Security (CCS), Alexandria, VA (October 2008)
9. Checkoway, S., Davi, L., Dmitrienko, A., Sadeghi, A.R., Shacham, H., Winandy, M.: Return-Oriented Programming without Returns. In: ACM Conference on Computer and Communications Security (CCS), Chicago, IL (October 2010)
10. Checkoway, S., Feldman, A.J., Kantor, B., Halderman, J.A., Felten, E.W., Shacham, H.: Can DREs Provide Long-Lasting Security? The Case of Return-Oriented Programming and the AVC Advantage. In: EVT/WOTE, Montreal, Canada (August 2009)
11. Chen, P., Xiao, H., Shen, X., Yin, X., Mao, B., Xie, L.: DROP: Detecting Return-Oriented Programming Malicious Code. In: Prakash, A., Sen Gupta, I. (eds.) ICISS 2009. LNCS, vol. 5905, pp. 163–177. Springer, Heidelberg (2009)
12. Chen, P., Xing, X., Han, H., Mao, B., Xie, L.: Efficient Detection of the Return-Oriented Programming Malicious Code. In: Jha, S., Mathuria, A. (eds.) ICISS 2010. LNCS, vol. 6503, pp. 140–155. Springer, Heidelberg (2010)
13. Chen, P., Xing, X., Mao, B., Xie, L., Shen, X., Yin, X.: Automatic construction of jump-oriented programming shellcode (on the x86). In: ASIACCS, Hong Kong, China (March 2011)
14. Chen, S., Xu, J., Sezer, E.C., Gauriar, P., Iyer, R.K.: Non-control-data attacks are realistic threats. In: USENIX Security Symposium, Baltimore, MD (July 2005)
15. Cifuentes, C., Van Emmerik, M.: UQBT: Adaptable Binary Translation at Low Cost. Computer 33(3), 60–66 (2000)
16. Cowan, C., Pu, C., Maier, D., Walpole, J., Bakke, P., Beattie, S., Grier, A., Wagle, P., Zhang, Q.: StackGuard: Automatic Adaptive Detection and Prevention of Buffer-Overflow Attacks. In: USENIX Security Symposium, San Antonio, TX (January 1998)

17. Davi, L., Dmitrienko, A., Nurnberger, S., Sadeghi, A.R.: Gadge Me If You Can: Secure and Efficient Ad-hoc Instruction-Level Randomization for x86 and ARM. In: ASIACCS, Hangzhou, China (May 2013)
18. Davi, L., Sadeghi, A.R., Winandy, M.: Dynamic integrity measurement and attestation: towards defense against return-oriented programming attacks. In: ACM Workshop on Scalable Trusted Computing (STC), Chicago, IL (November 2009)
19. Davi, L., Sadeghi, A.R., Winandy, M.: ROPdefender: A Detection Tool to Defend Against Return-Oriented Programming Attacks. In: ASIACCS, Hong Kong, China (March 2011)
20. Demay, J.C., Majorczyk, F., Totel, E., Tronel, F.: Detecting illegal system calls using a data-oriented detection model. In: International Information Security Conference (SEC), Lucerne, Switzerland (June 2011)
21. Dullien, T., Kornau, T., Weinman, R.P.: A framework for automated architecture-independent gadget search. In: USENIX Workshop on Offensive Technologies (WOOT), Washington, D.C. (August 2010)
22. Feng, H.H., Kolesnikov, O.M., Fogla, P., Lee, W., Gong, W.: Anomaly Detection Using Call Stack Information. In: IEEE Symposium on Security and Privacy, Oakland, CA (May 2003)
23. Francillon, A., Perito, D., Castelluccia, C.: Defending embedded systems against control flow attacks. In: ACM Workshop on Secure Execution of Untrusted Code (SecuCode), Chicago, IL (November 2009)
24. Fratric, I.: ropguard (2012), http://code.google.com/p/ropguard/
25. Gao, D., Reiter, M.K., Song, D.: Gray-box extraction of execution graphs for anomaly detection. In: ACM Conference on Computer and Communications Security (CCS), Washington, D.C. (October 2004)
26. Giffin, J.T., Jha, S., Miller, B.P.: Automated Discovery of Mimicry Attacks. In: Zamboni, D., Kruegel, C. (eds.) RAID 2006. LNCS, vol. 4219, pp. 41–60. Springer, Heidelberg (2006)
27. Harris, L.C., Miller, B.P.: Practical Analysis for Stripped Binary Code. ACM SIGARCH Computer Architecture News 33(5), 63–68 (2005)
28. Hiser, J., Nguyen-Tuong, A., Co, M., Hall, M., Davidson, J.W.: ILR: Where'd My Gadgets Go? In: IEEE Symposium on Security and Privacy, San Francisco, CA (May 2012)
29. Huang, Z., Zheng, T., Liu, J.: A Dynamic Detective Method against ROP Attack on ARM Platform. In: International Workshop on Software Engineering for Embedded Systems (SEES), Zurich, Switzerland (June 2012)
30. Huang, Z., Zheng, T., Shi, Y., Li, A.: A Dynamic Detection Method against ROP and JOP. In: International Conference on Systems and Informatics, Yantai, China (May 2012)
31. Kayaalp, M., Ozsoy, M., Abu-Ghazaleh, N., Ponomarev, D.: Branch Regulation: Low-Overhead Protection from Code Reuse Attacks. In: International Symposium on Computer Architecture (ISCA), Portland, OR (June 2012)
32. Kemerlis, V.P., Portokalidis, G., Keromytis, A.: KGuard: Lighweight Kernel Protection against Return-to-user Attacks. In: USENIX Security Symposium, Bellevue, WA (August 2012)
33. Kruegel, C., Kirda, E., Mutz, D., Robertson, W., Vigna, G.: Automating mimicry attacks using static binary analysis. In: USENIX Security Symposium, Baltimore, MD (July 2005)
34. Li, J., Wang, Z., Jiang, X., Grace, M., Bahram, S.: Defeating Return-Oriented Rootkits with "Return-Less" Kernels. In: European Conference on Computer Systems (EuroSys), Paris, France (April 2010)

35. Onarlioglu, K., Bilge, L., Lanzi, A., Balzarotti, D., Kirda, E.: G-Free: Defeating Return-Oriented Programming Through Gadget-less Binaries. In: Annual Computer Security Applications Conference (ACSAC), Austin, TX (December 2010)
36. Pappas, V., Polychronakis, M., Keromytis, A.D.: Smashing the Gadgets: Hindering Return-Oriented Programming Using In-Place Code Randomization. In: IEEE Symposium on Security and Privacy, San Francisco, CA (May 2012)
37. Paradyn Project: Dyninst (2013), http://www.dyninst.org
38. Paradyn Project: InstructionAPI (2013), http://www.dyninst.org
39. Paradyn Project: ParseAPI (2013), http://www.dyninst.org
40. Paradyn Project: ProcControlAPI (2013), http://www.dyninst.org
41. Paradyn Project: StackwalkerAPI (2013), http://www.dyninst.org
42. Parampalli, C., Sekar, R., Johnson, R.: A practical mimicry attack against powerful system-call monitors. In: ASIACCS, Tokyo, Japan (March 2008)
43. Polychronakis, M., Keromytis, A.: ROP Payload Detection Using Speculative Code Execution. In: International Conference on Malicious and Unwanted Software (MALWARE), Fajardo, Puerto Rico (October 2011)
44. Roemer, R., Buchanan, E., Shacham, H., Savage, S.: Return-Oriented Programming: Systems, Languages, and Applications. ACM Trans. Info. & Systems Security (TISSEC) 15(1), 2:1–2:34 (2012)
45. Rosenblum, N., Zhu, X., Miller, B., Hunt, K.: Learning to Analyze Binary Computer Code. In: AAAI, Chicago, IL (2008)
46. Roundy, K.A., Miller, B.P.: Hybrid Analysis and Control of Malware Binaries. In: Jha, S., Sommer, R., Kreibich, C. (eds.) RAID 2010. LNCS, vol. 6307, pp. 317–338. Springer, Heidelberg (2010)
47. Schwartz, E.J., Avgerinos, T., Brumley, D.: Q: Exploit Hardening Made Easy. In: USENIX Security Symposium. San Francisco, CA (August 2011)
48. Sekar, R., Bendre, M., Dhurjati, D., Bollineni, P.: A Fast Automaton-Based Method for Detecting Anomalous Program Behaviors. In: IEEE Symposium on Security and Privacy, Oakland, CA (May 2001)
49. Shacham, H.: The geometry of innocent flesh on the bone: return-into-libc without function calls (on the x86). In: ACM Conference on Computer and Communications Security (CCS), Alexandria, VA (October 2007)
50. sickness: Linux exploit development part 4 - ASCII armor bypass + return-to-plt (2011), http://sickness.tor.hu/?p=378
51. Szekeres, L., Payer, M., Wei, T., Song, D.: SoK: Eternal War in Memory. In: IEEE Symposium on Security and Privacy (May 2013)
52. Theiling, H.: Extracting safe and precise control flow from binaries. In: Conference on Real-Time Computing Systems and Applications, Washington, D.C. (December 2000)
53. Wartell, R., Mohan, V., Hamlen, K.W., Lin, Z.: Binary Stirring: Self-randomizing Instruction Addresses of Legacy x86 Binary Code. In: ACM Conference on Computer and Communications Security (CCS), Raleigh, NC (October 2012)
54. Zhang, C., Wei, T., Chen, Z., Duan, L., Szekeres, L., McCamant, S., Song, D., Zou, W.: Practical Control Flow Integrity & Randomization for Binary Executables. In: IEEE Symposium on Security and Privacy, San Francisco, CA (May 2013)
55. Zhang, M., Sekar, R.: Control Flow Integrity for COTS Binaries. In: USENIX Security Symposium, Washington, D.C. (August 2013)

Security@Runtime: A Flexible MDE Approach to Enforce Fine-grained Security Policies

Yehia Elrakaiby, Moussa Amrani, and Yves Le Traon

University of Luxembourg, 4 Alphonse Weicker L-2721, Luxembourg
{yehia.elrakaiby,moussa.amrani,yves.letraon}@uni.lu

Abstract. In this paper, we present a policy-based approach for automating the integration of security mechanisms into Java-based business applications. In particular, we introduce an expressive Domain Specific modeling Language (DSL), called *Security@Runtime*, for the specification of *security configurations* of targeted systems. The *Security@Runtime* DSL supports the expression of *authorization*, *obligation* and *reaction* policies, covering many of the security requirements of modern applications. Security requirements specified in security configurations are enforced using an *application-independent* Policy Enforcement Point (PEP)- Policy Decision Point (PDP) architecture, which enables the *runtime update* of security requirements. Our work is evaluated using two systems and its advantages and limitations are discussed.

Keywords: Java Security, Security Policies, Security Domain Specific Language, Access Control, Obligations.

1 Introduction

Integrating security mechanisms into applications is necessary to ensure data confidentiality, data integrity and users' privacy preservation. Security is a cross-cutting concern affecting most parts of an application and, therefore, decoupling security requirements from the code implementing system functionalities is desirable to achieve code modularity and simplify the correct development of systems and their maintenance. Previous works primarily focus on the separate specification of *access control requirements* and integration of access control enforcement mechanisms into applications using either *Aspect Oriented Programming* (AOP) [9] [4–8] or using a model-based approach [11–13]. In the former approach, access control enforcement mechanisms are automatically *weaved* into the application at compilation time, whereas in the latter approach, the system and its access control requirements are abstractly specified using *models*, from which implementation code is generated. Neither of these approaches allows for the *runtime* updating of security requirements.

The dynamic nature of modern applications and their sophistication requires however more than just *static* access control, typically the only security requirement covered in existing approaches (see Section 6 for a detailed discussion of

J. Jürjens, F. Piessens, and N. Bielova (Eds.): ESSoS 2014, LNCS 8364, pp. 19–34, 2014.

current approaches and their features). In particular, security requirements typically reflect regulatory and internal mandates, which are naturally dynamic and could change with time. Also, many systems today have requirements that go beyond access control such as usage control [1], which extends traditional access control by enabling specification of obligations that users must fulfill before, while or after access, and privacy obligations [3, 2], which dictate duties and expectations on how users' personal data should be handled.

In this paper, we propose a Domain Specific modeling Language (DSL) and an architecture for securing Java-based business applications to address the aforementioned issues. The DSL supports the expression of fine-grained contextual authorization, obligation, sanction and reaction policies, thus covering the expression of many of the sophisticated security requirements of modern applications. Security policies specified using the DSL are enforced into target applications using an application-independent architecture, which follows the Policy Enforcement Point (PEP) / Policy Decision Point (PDP) paradigm. The proposed architecture enforces security requirements into target applications in a non-intrusive manner using Aspect-Oriented Programming (AOP) [9], enabling a clean separation between the application's functional and non-functional requirements. Furthermore, the architecture supports the update of security requirements at runtime.

The remainder of the paper is organized as follows. Section 2 describes *S@R* (for *Security@Runtime*), our DSL for dealing with the identified challenges and its enforcement architecture. Section 3 illustrates our approach by presenting a complete application example. Section 4 describes the implementation of our tool prototype. Section 5 shows performance results of the prototype for two real-life systems. Section 6 presents related work; and Section 7 concludes the paper and discusses future work.

2   The Security@Runtime Approach

At the center of our approach is the *Security@Runtime (S@R)* DSL, used for the *configuration* of the security enforcement mechanisms. This Section starts by presenting our PDP/PEP architecture for enforcing *S@R* security *configurations*, then describes each *S@R* component in detail.

2.1   Architecture Overview

Figure 1 shows the main components of the security enforcement architecture, namely the (i) PAP, (ii) PEP and (iii) PDP. The Policy Administration Point (PAP) allows the specification of a *S@R configuration* for the PEP and the PDP. The PEP monitors the application, using AOP [9] (in our case, AspectJ), and filters out information that is irrelevant to policy enforcement based on the configuration. Three events are monitored by the PEP: *instance creation*, *instance field updates* and *method calls*. If an event is relevant to policy management or enforcement, then the PEP notifies the PDP to update the *effective* security policy accordingly, e.g. activate a new obligation. When an event corresponds to a

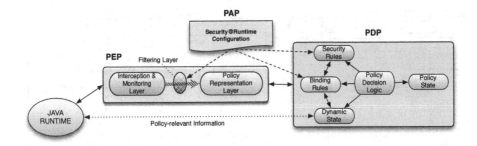

Fig. 1. Architecture Overview

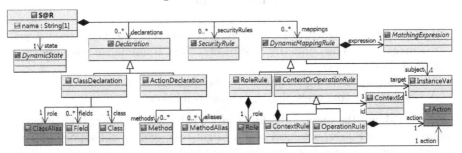

Fig. 2. The *S@R* Metamodel

method call, the PDP computes an access decision. If access is granted, then execution proceeds; otherwise, different actions are possible: (1) a runtime security exception is raised with an appropriate message, (2) the system is stopped, or (3) the method execution is skipped. In our current prototype's implementation, a security exception is raised after access denial.

Figure 2 shows the four building blocks of *S@R*: (1) DynamicState, (2) Declarations, (3) SecurityRules and (4) DynamicMappingRules. The *Dynamic State* is a partial representation of the runtime state of the application and is automatically maintained and managed by the PDP. The other blocks define the *configuration* of the security mechanisms and are presented successively in the following: SecurityRules are introduced in Section 2.2; then Declarations and DynamicMappingRules are described in Section 2.3. A comprehensive example is given in Section 3 to illustrate the specification of security configurations.

2.2 Security Rules (SR)

A security policy is a set of security rules specifying what *subjects*, i.e. active entities in the system, are permitted, prohibited and obliged to do in the system. A security rule is contextual, i.e. it may apply only under certain conditions. A security rule includes the following elements:

– An *Identifier* of the security rule.

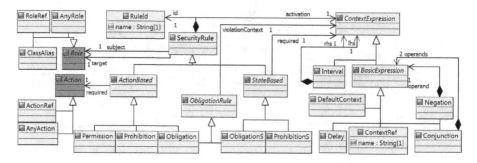

Fig. 3. Security Rules

- A *Role* representing a set of system users or resources (we choose to abstract resources using roles, similarly to subjects/users, to minimize the number of basic policy entities).
- An *Action* representing an interaction between users and resources.
- A *Context* denoting a set of system state conditions.

Figure 3 shows the metamodel for security rules. Each `SecurityRule` has a unique identifier RuleId, a subject and a target role, and an activation context. The activation context defines the rule's applicability condition: a rule is active only if the evaluation of the boolean ContextExpression is true. A security rule is either a Permission, a Prohibition, or an Obligation and may be either ActionBased or StateBased. An ActionBased rule specifies that its subject role is permitted, prohibited or obliged to execute an `Action` on its `target` role. A StateBased rule specifies that its subject role is obliged or prohibited to maintain a required `Context`. An Obligation defines a violationContext that specifies under which conditions the obligation, after its activation, should be considered violated.

A `ContextExpression` is a boolean expression language constituted of Basic-Expressions that can be composed with the usual boolean connectives. A Basic-Expression is either a DefaultContext, a ContextRef, or a Delay. A DefaultContext is a special context that is always true. A Delay is a context that is true after the elapse of the time period specified in it. Finally, a ContextExpression can be an `Interval`, denoted [lhr, rhs]. An interval context holds since the left-hand side `lhs` holds until the right-hand side `rhs` holds.

2.3 Declarations and Dynamic Security Rules

Security policies are defined on the abstract level on roles, actions and contexts, allowing the use of the same policies in different systems. Declarations and DynamicMappingRules link elements of security rules to target applications by defining a mapping between these elements and the application classes, instances and method calls.

Declarations define aliases for the application classes and methods to simplify referring to them in security rules instead of using fully qualified names. A

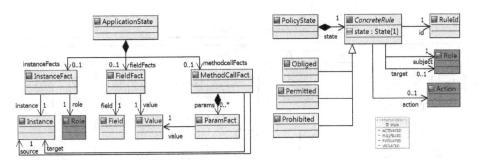

Fig. 4. Dynamic State

Declaration (cf. Fig. 2) is either a `ClassDeclaration` or an `ActionDecla-`
`ration`. A `ClassDeclaration` provides an alias for one application class and
may optionally specify a list of the `Fields` of the class that are relevant to the
enforcement of the security policy. This list improves system efficiency: only the
updates of the relevant instance fields will be notified to the PDP, as opposed to
notifying the PDP about changes of the value of every field of declared classes (see
implementation details in Section 4). An `ActionDeclaration` provides an alias
for one of the application methods or for every method in a sequence of (nested)
methods. Declarations indicate which parts of the application are *relevant* to the
security policy, therefore they are used by the PEP to filter information about
changes in the application state that are being notified to the PDP.

Dynamic Mapping Rules describe the mapping between the *policy* entities (roles, actions and contexts) and the *application* entities (instances, fields,
methods and their parameters). A RoleRule specifies which instances in the application are assigned to which role in the policy. An OperationRule specifies a
correspondence between method calls and policy actions. A ContextRule defines
a policy context as a condition on the application state: the context holds if its
MatchingExpression holds on the application state. In the following, we describe
the representation of the *application* and *policy* states (which compose together
the PDP state) in *S@R*. Then, we explain how elements of the application state
are mapped to elements of the policy state using dynamic mapping rules.

Application State At runtime, the state of an application consists of the set of
active objects (or instances), the field instance values, and the stack of method
calls. To correctly manage the security policy, e.g. activate contextual obligations, changes in the application state that are relevant to the enforcement of
the policy need to be monitored. In our architecture, shown in Figure 1, these
changes are monitored by the aspect layer within the PEP, which notifies the
PDP when a relevant change is detected. Using the PEP's notifications, the PDP
maintains a partial representation of the application state. Concretely, this state
takes the form of a set of First-Order Logic (FOL) facts, allowing the specification of security policies in FOL. This state is metamodelled in Figure 4 (left): an
InstanceFact represents a class instance; a FieldFact represents an instance field;

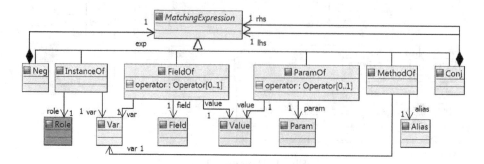

Fig. 5. Rule Condition Language

a MethodCallFact, together with ParamFacts, represents a method call. In *S@R*'s concrete syntax, these facts are written as follows:

- `instance_of(i,r)`: i is an instance of class r.
- `field_of(i,f,v)`: v is the value of the field f of the instance i.
- `call_of(c,m)`: c is a call of the method m.
- `param_of(c,p,v)`: v is the value of parameter p in the method call c.

Two special parameters, this and target, are systematically added to the normal parameter list of a method call to denote the calling and called instance respectively.

Policy State is managed by the PDP based on the application state. A PolicyState contains security rules that are *applicable*, or *effective* at a given time. Effective security rules are the set of ACTIVE permissions and the set of ACTIVE, FULFILLED or VIOLATED obligations (all values of the State enumeration in Fig. 4). In *S@R*'s concrete syntax, a policy state is represented using facts having one of the following forms:

- `permitted(r,s,a,o)`: rule r authorizes subject s to take action a on o.
- `prohibited(r,s,a,o)`: rule r prohibits s to take a on o.
- `obliged(r,s,a,o,t)`: rule r obliges s to take a on o.
- `obliged(r,s,c,o,t)`: rule r obliges s to maintain c on o.

Dynamic Mapping Rules If the MatchingExpression of a DynamicMappingRule holds on the application and policy states for some instantiation, i.e. a variable substitution making the MatchingExpression true, then the DynamicMappingRule holds for this instantiation. For example, a RoleRule of the form `role_of(I, role_name) <- instance_of(I,class_name)` assigns an instance x to the role role_name when `instance_of(x,class_name)` holds in the current state. Figure 5 shows the MatchingExpression's metamodel: it includes a matching expression for matching a class instance, a field value, a method call, or any logical combination of these elements. Note that data operators can be used for the FieldOf and ParamOf expressions. For example, `field_of(D,age,≤,18)` means that D is any instance whose value for the field age is less or equal than 18.

Table 1. Security Requirements for Hospital X

R1 A doctor can read the medical files of the patients he's treating [ACCESS CONTROL]
R2 A doctor should fill and submit a case evaluation report for each of his patients within one week [NON-PERSISTENT OBLIGATION]
R3 A doctor should fill and submit a check-up report when he is assigned a patient within two days [PERSISTENT OBLIGATION]
R4 If the doctor does not submit the report, then he has to fill and submit a "violation of duty" report within one week [SANCTIONS]
R5 Meanwhile, his access to all files are suspended [REACTIONS]
R6 A doctor has to delete his patients' files that are private and not vital within two years [PRIVACY]
R7 A medical file has to be stored encrypted at most one minute after its creation by its creator [CONFIDENTIALITY]

3   Example: The Medical System (MS)

Consider as an example the information system of Hospital X. The hospital needs to comply with some internal and regulatory mandates governing the activities of its personnel in order to protect the privacy of patients and guarantee the confidentiality and integrity of its information system. In Hospital X, a security policy should be specified to govern interactions between Doctors, Patients, Reports and Files, each of these roles being implemented into a simple class in the application. Table 1 describes this policy informally. The policy specifies one access control requirement (R1), a *non-persistent obligations* (R2), i.e. an obligation that may be cancelled after it is activated, *persistent obligations* (R3), i.e. an obligation that cannot be cancelled, *sanctions* (R4) and *reactions* (R5) to compensate the violation of obligations, and the obligations R6 and R7 to satisfy some privacy and confidentiality requirements respectively.

To enforce these requirements, the security officer of Hospital X should define a security configuration for the information system of Hospital X as follows: (i) *declare* the monitored aspects of the information system using Declarations; (ii) specify how application entities are *mapped* to policy entities using DynamicMappingRules; and (iii) define the SecurityRules formalizing the regulatory mandates.

Declarations are simply specified by defining aliases for classes and methods that need to be referenced in other parts of the configuration (dynamic mapping rules and security rules). The following example creates an alias "doctor" for the class *.Person.Doctor, declaring its field patients as the only policy relevant field (when the **attribute** clause is absent, all fields are considered relevant). Note that class aliases can then be used directly as role names (cf. Fig. 3). Similarly, an alias "read" is created for the method *.Server.readFile(String). Finally, the actions "readServer" and "readFile" are used as aliases for the method calls appearing in the sequence (denoted by ->) of the method calls readFileServer followed by readFile.

class_alias	doctor	method_id	read
class	*.Person.Doctor	method_sig	*.Server.readFile(String)
attributes	ArrayList<Patient>:	method_id	readServer,readFile
	patients	method_sig	*.Server.readFileServer(String)
			->*.File.readFile()

Security Rules are defined according to the Security Requirements (cf. Tab. 1). Here, each requirement is expressed using one security rule. Rule r3 has doctor, submit and report as its subject, action and target respectively. The rule's activation context is an Interval, thus r3 is activated when the context assigned_doctor becomes true, and is never cancelled (because the Interval's rhs is false). Rule r2 is non-persistent because its activation context is a BasicExpression: in this case, the obligation is activated when assigned_doctor holds, i.e. when a doctor is assigned to some patient, and it is cancelled when the activation context no longer holds, i.e. if this patient is no longer treated by this doctor.

```
permission(r1,doctor,read,file, assigned_doctor)
action_obl(r2,doctor,submit,report,assigned_doctor,delay<1:w>)
action_obl(r3,doctor,submit,report,[assigned_doctor,false],delay<2:d>)
action_obl(r4,doctor,submit,viol_report,violation_r2,delay<1:w>)
prohibition(r5,doctor,read,file,violation_r2)
action_obl(r6,doctor,delete,file,private & !vital, delay<2:y>)
state_obl(r7,doctor,file_encrypted,file,file_created,delay<1:m>)
```

Dynamic Mapping Rules are defined as follows. The first mapping rule is a RoleRule: it says that any instance of the class report whose field type has the value of 'violation report' is assigned to the role of viol_report. The second mapping rule is an OperationRule: it says that if an instance (denoted here by S) calls the method read_method on a file (denoted by F), then the policy action "read" has subject S and target F (note the use of the special parameters this and target for the calling/callee instances). The fourth rule is a ContextRule that specifies that the context private holds for any file F whose field classification has the value of ''private''. The ContextRule violation_r2 is different, because it depends on the policy state: violation_r2 holds for a doctor D for which r2 is in the VIOLATED state.

role_of(R,viol_report)	← instance_of(R,report) & field_of(R,type,'violation report')
operation(S,read,F)	← call_of(read_method) & param_of(read_method,this,S) & param_of(read_method,target,F)
operation(S,delete,F)	← call_of(delete_method) & param_of(delete_method,this,S) & param_of(delete_method,target,F)
hold(_,_,F,private)	← instance_of(F,file) & field_of(F,classification,'private')
hold(D,_,_,violation_r2)	← violated(r2,D,submit,report) field_of(F,type,'vital')
hold(D,_,_,file_created)	← instance_of(F,file) & field_of(F,creator,D) & instance_of(D,doctor)
hold(D,_,P,assigned_doctor)	← instance_of(D,doctor) & field_of(D,patients,contains,P) & instance_of(P,patient)
hold(_,_,F,file_encrypted)	← instance_of(F,file) & field_of(F,encrypted,true)

One could also define the action **read** differently: an instance S **reads** F whenever
readServer and **ReadFile** are called sequentially; then S is matched to the caller
instance of the method aliased to **readServer** (precisely, *.Server.readFile-
Server(String), as declared before) and F to the target instance for the method
aliased to **readFile** (declared previously as being *.File.readFile()).

```
operation(S,read,F) ← call_of(read_server) & call_of(read_file) &
                      param_of(read_server,this,S) & param_of(read_file,target,F)
```

---

4   Implementation

The architecture described in Fig. 1 is implemented using AspectJ for monitoring
the target application, XSB Prolog [34] for computing access control decisions
and policy management and Java/interProlog [35] for the communication between the PEP and the PDP. EMFText [33] is used for parsing *S@R*'s concrete
syntax and creating models.

4.1   Application Monitoring Layer

Each activity affecting the application state (instance creation, field update
or method call) is monitored using an aspect. When an instance is created,
if its class type is part of the Declarations within the *S@R* configuration, the
aspect **RelevantClassObserver** detects the object using a pointcut of the form
execution(*.new (..)), and passes it to the representation layer.

Field value update detection is more sophisticated: pointcuts of the form
set(* *) only works for Java primitive types (strings, integers, booleans and
so on). To monitor changes within the other supported data structures (like
ArrayLists), a specific pointcut is defined to detect the execution of all methods altering the contents of the data structure (for example, ArrayList.clear
or ArrayList.set). The pointcut below is specifically defined for the class
ArrayList. Currently, our implementation supports fields whose type is a primitive type and unidimensional structures (Vectors, HashSets and ArrayList).
It is however straightforward to support more data structures.

Finally, method calls are intercepted using a pointcut of the form (call(public
* *(..))). This aspect is of type *around,* i.e. the call is not executed until the
aspect code is executed. This allows verification of the policy state at the PDP
before allowing the execution of the method.

```
protected pointcut arrayListUpdate():
    (call(* java.util.ArrayList.add*(..)) ||
         call(* java.util.ArrayList.clear(..)) ||
         call(* java.util.ArrayList.remove*(..)) ||
         call(* java.util.ArrayList.retain*(..)) ||
         call(* java.util.ArrayList.set*(..))
         )
         && within(*.*) && !within(sr.*);
```

4.2   Policy Representation Layer

The policy representation layer consists of a recursive algorithm that processes
Java objects and method calls in order to represent them using the facts de-
scribed in Figure 4. For example, consider an instance of the declared class
doctor with a field age of type Integer and another field patients of type
ArrayList of Patient. Class Patient has a single field name of type String.
Suppose there are two patients P_1 and P_2 in the treated patients list of doctor
X. These objects would be represented as follows:

- instance_of(X,doctor) representing the instance,
- field_of(X,age,18) if the value of age of X is 18,
- field_of(X,patients,Y) where Y is an identifier of the ArrayList of patients,
- field_of(Y,e,P_1) where P_1 is an identifier of the first Patient, this fact denotes
  that P_1 is an element of the ArrayList of patients,
- field_of(Y,e,P_2) where P_2 is an identifier of the second Patient,
- field_of(P_1,name,'John') if the name of the first Patient is John,
- field_of(P_2,name,'Ben') if the name of the second Patient is Ben.

A method call is represented using a fact of the form call_of(method_id) where
method_id is the method alias. Methods are processed similarly to class instances. However, param_of facts use the parameter position instead of attribute
names: for example, a fact param_of(delete_method,1,X) means that X is the first
parameter of the method call delete_method. Method calls have two additional
parameters, namely the this and target parameters denoting the calling and called
instances of the method call respectively. AspectJ enables an easy identification
of these instances for intercepted method calls.

4.3   The Policy Decision Point (PDP)

The PDP is a policy engine implemented in Prolog. It computes a decision for
access control requests and manages obligations in the policy according to notifications received from the PEP as follows: for an instance creation notification,
new Prolog facts representing the instance are inserted into the engine's knowledge base; for a field update notification, old facts specifying the field's value
are retracted and replaced by new facts specifying the new value; for a method
call notification, new facts corresponding to the call and its parameters' value
(which works just as for instance fields) are inserted and the authorization policy
is checked. An access control decision is then returned to the PEP and the facts
corresponding to the call are retracted from the engine. After each notification,
the PDP also updates the state of obligations.

Access Control After a method call is attempted, the call is interpreted by
the PEP. An access is granted if it is permitted and not prohibited, i.e. the
prohibitions are given an implicit priority over permissions for the resolution of
potential conflicts between them. This access control policy evaluation strategy
is specified using Prolog rules as follows:

Table 2. Obligation Management Rules

	Action	Conditions
1	assert(obliged(I,S,A,T,[Ca,Cd],Cv,active))	action_obligation(I,R,A,Rt,[Ca,Cd],Cv) instance_of(S,R) instance_of(T,Rt) hold(S,A,T,Ca)
2	retract(obliged(I,S,A,T,[Ca,Cd],Cv,active)) assert(obliged(I,S,A,T,[Ca,Cd],Cv,violated))	obliged(I,S,A,T,[Ca,Cd],Cv,active) hold(S,A,T,Cv)
3	retract(obliged(I,S,A,T,[Ca,Cd],Cv,active)) assert(obliged(I,S,A,T,[Ca,Cd],Cv,fulfilled))	obliged(I,S,A,T,[Ca,Cd],Cv,active) operation(S,A,T)
4	retract(obliged(I,S,A,T,[Ca,Cd],Cv,_))	obliged(I,S,A,T,[Ca,Cd],Cv,_) hold(S,A,T,Cd)
5	retract(obliged(I,S,Cf,T,[Ca,Cd],Cv,active)) assert(obliged(I,S,Cf,T,[Ca,Cd],Cv,fulfilled))	obliged(I,S,Cf,T,[Ca,Cd],Cv,active) hold(S,_,T,Cf)

```
allow_operation(S,A,T) ← operation(S,A,T), permitted(_,S,A,T), ¬ prohibited(_,S,A,T).

permitted(I,S,A,T)     ← permission(I,Rₛ,A,Rₜ,C), instance_of(S,Rₛ), action(A),
                         instance_of(T,Rₜ), hold(S,A,T,C).

prohibited(I,S,A,T)    ← prohibition(I,Rₛ,A,Rₜ,C), instance_of(S,Rₛ), action(A),
                         instance_of(T,Rₜ), hold(S,A,T,C).
```

Obligations The PDP manages the state of obligations by detecting their activation, cancellation, fulfillment and violation. We unify the representation of obligation activation contexts using Intervals: every activation context c is represented using an interval [c, !c].

An obligation is managed as follows: it is instantiated when its activation context holds. It is then fulfilled when its required action (context) is detected, otherwise the obligation is violated if its violation context holds. An obligation is canceled at any time when its deactivation context holds. Table 2 shows the conditions for the detection of the activation, cancellation, fulfillment and violation of obligations and the update actions taken by the PDP when these conditions are detected. For example, Line 3 specifies that when the state of an obligation is active, its state is updated to violated when its violation context becomes true. State obligations (Line 5) are managed similarly, however their fulfillment is detected when their required context holds.

Support of Quaternary Predicates One advantage of using predicate logic to represent the application state is that it simplifies the definition of predicates for the expression of sophisticated state conditions. For example, the rules below specify the *contains* operator for data structures like *ArrayLists*, the less than or equals operator for numbers and the derivation of *violated* facts from obligation facts respectively.

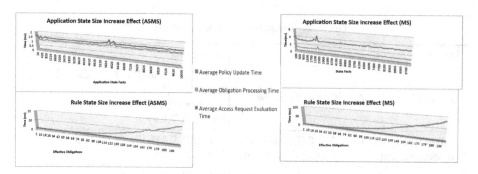

Fig. 6. Performance Results

```
field_of(Id,Name,<=,Val)        ← field_of(Id,Name,V), number(V), V =< Val.

field_of(Id,Name,includes,Val)  ← field_of(Id,Name,V), field_of(V,e,Val).

violated(Id,S,A,O)              ← obliged(Id,S,A,O,[C_a,C_d],C_v,violated).
```

4.4 Policy Update

A runtime update of Security Rules is managed by simply adding (or retracting) the Security Rules from the policy engine (and their associated facts), i.e. if an obligation rule is removed then all its activated instances are also removed. When a new class is added (or removed) to the declarations part of the configuration, then the instances of this class are added to (or removed from) the knowledge base of the PDP as discussed in Section 4.2. Similarly, when an attribute is added (or removed), then the facts representing it are added (or retracted). A method call that is declared (undeclared) starts (ceases) to be monitored by the PEP. To handle these updates to declared classes and attributes, we need to keep a map of the class instances of the application at the PEP level.

5 Validation

To validate our approach, we considered two use case applications, namely our MS running example of Section 3, and an Auction Sales Management System (ASMS). The ASMS consists of 122 classes, 797 methods and about 11 kLoC. The ASMS implements an Auction system, where users can buy or sell products online, after joining an auction and placing bids. Users can also post, or read, comments from the Auction session. In the evaluation, we specifically targeted performance-related research questions:

- does the tool perform sufficiently well to be used in practice?
- what are the main factors impacting the tool performance?

To answer these questions, we defined a security policy for each example application and ran a scenario covering the different policy management operations,

Table 3. Comparison of *S@R* with Existing MDS approaches

	DM	SM	CS	SC	CG
SecureUML[19]	UML	Profiles	OCL	AC	✓
UMLsec[20]	UML	Profiles	×	C,IF,AC	×
secureMDD[21]	UML	Profiles	×	AC	✓
ModelSec[31]	UML	SecML	∘	MC	✓
SECTET[32]	SE-UML	Profiles	SE-PL	AC	∘
XACML[36]	×	XACML	XACML	AC,OB	×
S@R	UML	S@R	S@R	AC,OB	✓

AC: Access Control, C: Confidentiality,
IF: Information Flow, OB: Obligations, MC: Multiple Concerns

DM: System Modeling, **SM**: Security Modeling, **CL**: Contextual Security
SC: Security Concerns, **CG**: Code Generation

i.e. obligation activation, violation, access control, etc. We evaluated three factors: the time necessary to perform policy management operations, to evaluate an access request, and to update the (obligation) policy state, for different sizes of the application and policy states (number of activated obligations).

Figure 6 shows the results: policy management operations and access request evaluations are performed in a few milliseconds, and represent an almost constant overhead. On the other hand, obligation processing time increases with the number of (activated) obligations in the system: the activated obligations' contexts have to be verified individually to check whether they are canceled, violated or fulfilled, after each state update. We are currently investigating ways to improve this processing time.

6   Discussion and Related Work

Since the seminal contributions of Lodderstedt and Basin with SecureUML [10], and Jürgens with UMLSec [20] back in 2002, model-based development of secure systems has been an active research area. In Table 3, we compare several contributions with respect to several dimensions: system (DM) and security modeling (SM); contextual security (CS); security concerns (SC) (i.e. what kind of security properties can be expressed); and code generation (CG).

Domain and Security Modeling. UML is the most common way to define the target application domain, as shown in the first column of Table 3. UML-based approaches annotate the business UML model with their security requirements. Conceptually, our approach is different since we introduce the *S@R* DSL to specify security requirements and their mapping to target systems. One advantage of our approach is that it cleanly separates security from system specification making a true separation of concerns, as opposed to the use of OCL constraints to specify contextual policies when UML profiles are used. Note that XACML does not assume any specific domain modeling language and, therefore, it does not provide means to systematically integrate security mechanisms for enforcing XACML policies in targeted systems.

Expressivity of Security Languages is a major challenge since it is necessary to cover the specification of many practical security concerns. Many systems today have security requirements that go beyond access control, as recognized by Basin, Clavel and Egea in [13] where they pointed out the need to add support for obligations. To the best of our knowledge, S@R is the first DSL that supports management and enforcement of both authorizations and obligations. The specification of obligations is supported in XACML [36]. However, obligations in XACML are syntactic elements without formal semantics. Furthermore, XACML does not provide management and enforcement support for obligations.

Violation Monitoring and Policy Runtime Updating. Runtime policy updating and security rule violation monitoring are not, to the best of our knowledge, present in any of the current approaches.

Security Infrastructure. Despite the use of automated transformations in MDS, it is still difficult to enable full code generation from high-level requirements. SecureUML [19] supports the generation of secure systems for two target architectures (Enterprise JavaBeans and Microsoft DotNet), but the generating mechanism relies on pre-existing security mechanisms. In SECTET, the information relevant for authorizations are specified using the tool's language, and transformed into XACML specifications. In [31], security and business are composed into a model from which Java code is generated. Our approach generates Prolog code from security policies defined within S@R, whereas the rest (i.e., aspects monitoring and policy interpretation) is application-independent.

7   Conclusion

This paper proposes an approach for securing Java-based business applications using security policies. Our approach cleanly separates between security and business concerns, allowing the separate development and specification of business and security aspects. It also enables the specification of fine-grained contextual permissions and obligations and supports their management, enforcement and their update at runtime. We have demonstrated the expressiveness of our security policy language using a comprehensive example and validated our approach by using it to secure two different systems. We have identified some limitations of our framework, namely its scalability when the number of activated obligations in the system increases. Therefore, we plan to study optimization techniques to improve the tool's performance. We also intend to provide support for more advanced usage controls and more Java data structures.

References

1. Sandhu, R., Park, J.: The UCON ABC usage control model. ACM Transactions on Information and System Security (TISSEC) 7(1), 128–174 (2004)
2. Ni, Q., Bertino, E., Lobo, J.: An obligation model bridging access control policies and privacy policies. In: SACMAT 2008, p. 133 (2008)
3. Mont, M.: Dealing with privacy obligations in enterprises. In: ISSE 2004 Securing Electronic Business Processes, pp. 28–30 (2004)
4. Erlingsson, U., Schneider, F.B.: SASI enforcement of security policies. In: NSPW, pp. 87–95 (2000)
5. Bauer, L., Ligatti, J., Walker, D.: Composing security policies with polymer. ACM SIGPLAN Notices 40(6), 305 (2005)
6. de Oliveira, A.S., Wang, E.K., Kirchner, C., Kirchner, H.: Weaving rewrite-based access control policies. In: FMSE, pp. 71–80 (2007)
7. Hamlen, K.W., Jones, M.: Aspect-oriented in-lined reference monitors. In: PLAS, p. 11 (2008)
8. Hussein, S., Meredith, P., Rolu, G.: Security-policy monitoring and enforcement with JavaMOP. In: PLAS, pp. 1–11 (2012)
9. Kiczales, G., Lamping, J., Mendhekar, A., Maeda, C., Lopes, C.V., Loingtier, J.M., Irwin, J.: Aspect-Oriented Programming. In: Akşit, M., Matsuoka, S. (eds.) ECOOP 1997. LNCS, vol. 1241, pp. 220–242. Springer, Heidelberg (1997)
10. Lodderstedt, T., Basin, D.: SecureUML: A UML-Based Modeling Language for Model-Driven Security. In: Proceedings of the 5th International Conference on The Unified Modeling Language, pp. 426–441 (2002)
11. Mouelhi, T., Fleurey, F., Baudry, B., Le Traon, Y.: A model-based framework for security policy specification, deployment and testing. In: Czarnecki, K., Ober, I., Bruel, J.-M., Uhl, A., Völter, M. (eds.) MODELS 2008. LNCS, vol. 5301, pp. 537–552. Springer, Heidelberg (2008)
12. Morin, B., Mouelhi, T., Fleurey, F., Le Traon, Y., Barais, O., Jézéquel, J.M.: Security-driven model-based dynamic adaptation. In: ASE 2010 (2010)
13. Basin, D., Clavel, M., Egea, M.: A decade of model-driven security. In: SACMAT 2011, pp. 1–10 (2011)
14. Basin, D., Clavel, M., Doser, J., Egea, M.: A Metamodel-Based Approach for Analyzing Security-Design Models. In: Engels, G., Opdyke, B., Schmidt, D.C., Weil, F. (eds.) MODELS 2007. LNCS, vol. 4735, pp. 420–435. Springer, Heidelberg (2007)
15. May, M., Gunter, C., Lee, I.: Privacy APIs: Access control techniques to analyze and verify legal privacy policies. In: 19th IEEE Computer Security Foundations Workshop, CSFW 2006 (2006)
16. Barth, A., Datta, A., Mitchell, J., Nissenbaum, H.: Privacy and contextual integrity: framework and applications. In: IEEE Symposium on Security and Privacy (2006)
17. Barth, A., Mitchell, J., Datta, A., Sundaram, S.: Privacy and Utility in Business Processes. In: 20th IEEE Computer Security Foundations Symposium, pp. 279–294 (2007)
18. Lam, P.E., Mitchell, J.C., Sundaram, S.: A formalization of HIPAA for a medical messaging system. In: Fischer-Hübner, S., Lambrinoudakis, C., Pernul, G. (eds.) TrustBus 2009. LNCS, vol. 5695, pp. 73–85. Springer, Heidelberg (2009)
19. Basin, D., Doser, J., Lodderstedt, T.: Model driven security: From UML models to access control infrastructures. ACM Transactions on Software Engineering and Methodology (TOSEM) 15(1), 39–91 (2006)

20. Jürjens, J.: UMLsec: Extending UML for secure systems development. In: Jézéquel, J.-M., Hussmann, H., Cook, S. (eds.) UML 2002. LNCS, vol. 2460, pp. 412–425. Springer, Heidelberg (2002)
21. Moebius, N., Stenzel, K., Grandy, H., Reif, W.: SecureMDD: a model-driven development method for secure smart card applications. In: International Conference on Availability, Reliability and Security, ARES 2009, pp. 841–846 (March 2009)
22. Cuppens, F., Miège, A.: Modelling contexts in the Or-BAC model. In: ACSAC, pp. 416–425 (2003)
23. Elrakaiby, Y., Cuppens, F., Cuppens-Boulahia, N.: Formal enforcement and management of obligation policies. In: Data & Knowledge Engineering, pp. 1–21 (2011)
24. Jajodia, S., Samarati, P., Subrahmanian, V.: A logical language for expressing authorizations. In: Proceedings of 1997 IEEE Symposium on Security and Privacy, pp. 31–42 (1997)
25. Kagal, L., Finin, T.: A policy language for a pervasive computing environment. In: IEEE 4th International Workshop on Policies for Distributed Systems and Networks, pp. 63–74 (2003)
26. Gosling, J., Joy, B., Steele, G., Bracha, G., Buckley, A.: The Java Language Specification. Addison-Wesley Longman (2013)
27. Ben-Ghorbel-Talbi, M., Cuppens, F., Cuppens-Boulahia, N., Bouhoula, A.: A delegation model for extended RBAC. International Journal of Information Security 9(3), 209–236 (2010)
28. Cuppens, F., Cuppens-Boulahia, N., Ghorbel, M.B.: High Level Conflict Management Strategies in Advanced Access Control Models. Electronic Notes in Theoretical Computer Science 186, 3–26 (2007)
29. Autrel, F., Cuppens, F., Cuppens-Boulahia, N., Coma, C.: Motorbac 2: a security policy tool. In: 3rd Conference on Security in Network Architectures and Information Systems (SAR-SSI 2008), Loctudy, France, pp. 273–288 (2008)
30. Kateb, D.E., Mouelhi, T., Traon, Y.L., Hwang, J., Xie, T.: Refactoring access control policies for performance improvement. In: ICPE, pp. 323–334 (2012)
31. Molina, F., Toval, A., Sánchez, O., Garca-Molina, J.: ModelSec: A Generative Architecture for Model-Driven Security. Journal of Universal Computer Science 15(15), 2957–2980 (2009)
32. Breu, R., Popp, G., Alam, M.: Model based development of access policies. International Journal on Software Tools for Technology Transfer 9(5-6), 457–470 (2007)
33. emfText, http://www.emftext.org/index.php/EMFText
34. XSB Porlog, http://xsb.sourceforge.net
35. interProlog, http://www.declarativa.com/interprolog
36. Extensible Access Control Markup Language (XACML) version 3.0, http://docs.oasis-open.org/xacml/3.0/xacml-3.0-core-spec-cs-01-en.pdf

Idea: Towards a Vision of Engineering Controlled Interaction Execution for Information Services*

Joachim Biskup and Cornelia Tadros

Fakultät für Informatik, Technische Universität Dortmund, Germany
{joachim.biskup,cornelia.tadros}@cs.tu-dortmund.de

Abstract. To protect an agent's own knowledge or belief against unwanted information inferences by cooperating agents, Controlled Interaction Execution offers a variety of control methods to confine the information content of outgoing interaction data according to agent-specific confidentiality policies, assumptions and reaction specifications. Based on preliminary experiences with a prototype implementation as a frontend to a relational DBMS, in this article we outline the architectural design and the parameterized construction of specific tasks to uniformly shield all information services in need of confinement, potentially comprising query answering, update processing with refreshments, belief revision, data publishing and data mining. Refraining from any intervention at the cooperating agents, which are also seen as intelligently attacking the defending agent's own interest in preserving confidentiality, the engineering solely aims at self-confinement when releasing information.

Keywords: agent, a priori knowledge, attacker assumption, belief, belief revision, confidentiality policy, constraint, censor, data mining, data publishing, formal semantics, frontend, group, inference control, inference-usability confinement, information engineering, information flow, information integration, invariant, logic, lying, overestimation, permission, possibilistic secrecy, prohibition, query answering, reasoning, refreshment, refusal, simulation, state, theorem-proving, update processing.

1 Introduction

People are communicating by using their computing devices – profiting from external facilities while purposely either sharing or protecting own informational resources. We consider the people's devices as a kind of intelligent agents which are interacting within a multiagent system. Accordingly, while being designed to cooperatively share its data with another agent in general, each agent also has to keep its own sensitive information confidential. In this report, we outline a specific approach to engineer such a "defending" agent, for the sake of supporting privacy as informational self-determination and protecting business assets.

---

* This work has been supported by the Deutsche Forschungsgemeinschaft (German Research Council) under grant SFB 876/A5 within the framework of the Collaborative Research Center "Providing Information by Resource-Constrained Data Analysis".

J. Jürjens, F. Piessens, and N. Bielova (Eds.): ESSoS 2014, LNCS 8364, pp. 35–44, 2014.

Our approach can be motivated by the case of Bob living with his family and running his own business, gathering any information he needs and maintaining the plans he pursues by means of his computing devices. This includes XML documents with personal details about him and his children, their electronic health records, a relational database about commercial offers and customers, and an AI system for assembling and evaluating ideas on further projects.

Bob is cooperating with a large variety of relatives, friends, business partners, officials and so on. Communicating with any of them, according to the agreed purposes behind the specific contact, he is willing to discretionarily share some pieces of information and selected details of his plans. At the same time, however, depending on the social relationship and guided by personal preferences, Bob might want each of the persons involved not to learn parts of the information and plans that appear to be sensitive regarding the specific situation.

So, Bob is facing the challenge to uniformly and consistently shield the information services he offers to others, balancing his and their interests in availability and integrity of data to be shared with his potentially conflicting own interest of context-specific confidentiality. Accordingly, he would like to employ a single control mechanism to confine the outgoing information flow from all his computing devices according to his personal or business needs and the nature of the social contact to the recipient envisioned. To be sure, this mechanism should not at all be invasive to the computing devices of the others, but only regulate the functionality of his own devices.

In other words, Bob seeks for installing and parameterizing a personal intelligent computing agent that securely mediates the interactions with the devices of the others such that the specifically expressed interests are actually automatically enforced, in each single interaction and over the time as well.

More abstractly, we aim at constructing a *defending agent*, i.e., an intelligent computing agent that will enable its owner to both conveniently and effectively deal with formal requirements about the following high-level issues:

- general *permission* to share data, for the sake of *availability*: declared as *interface* language;
- dedicated *prohibition* to acquire information, for the sake of *confidentiality*: declared as a *policy*, expressing *security constraints/invariants* on "released information";
- application-oriented *quality guarantees* to reflect aspects of the "real-world", for the sake of *integrity*: declared as *functional constraints/invariants* on own information.

2   From a Required Vision to Available Ideas

Our *vision* of a defending agent is driven by widespread requirements to enable individuals to discretionarily control the sharing of their information with others, which are treated as cooperation partners on the one hand and nevertheless perceived as potential attackers against wanted confidentiality on the other hand. While the requirements appear to be socially accepted, actually offered

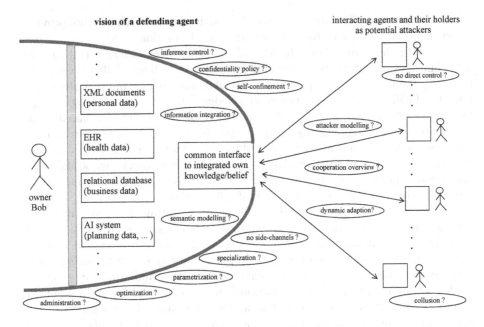

Fig. 1. The vision and major challenges

IT-systems rarely comply with them, and some people might even consider our vision to be just a dream, nice to have but impossible to achieve. We deal with the latter concern in a gradual way, rather than giving a strict yes-no answer.

In fact, in this article we argue for the thesis that our vision is approximately realizable indeed, by exhibiting an architectural design that exploits or partly is justified by *ideas* about actually *available technologies*. Trying a somehow bold correspondence, realizing the vision of controlling the release of own information is comparable to traveling to the moon: having been a dream for a long time, at a specific stage of scientific development, a concrete top-down plan was possible based on ideas from various fields. Clearly, just reaching the moon has left many further issues open, and so will our design be in need of many further features.

Figure 1 illustrates the vision and indicates major challenges. In the following, we gather most of these challenges into three main groups, and for each group we identify a basic idea about available technologies to solve the problems involved.

The challenges of inference control according to a confidentiality policy by self-confinement of the information supplier include the problems of attacker modelling and a complete cooperation overview. As a main idea towards a solution, we follow the approach of *Controlled Interaction Execution*, as summarized in Section 3. Though this approach is mainly logic-oriented, additional features are expected to be smoothly integrable, in particular numerical information like outputs of statistical functions and data mining support and confidence numbers. Attacker modelling in a large scale and, in particular, establishing an overview about cooperations are in need of adapting insight from, e.g., adversarial reasoning [18] and normative reasoning [1] about an inaccessible environment which does not provide complete knowledge about the dissemination of information.

We can further profit from the idea of *information engineering* [14] to deal with information integration requiring semantic modelling of diverse sources including their dynamic adaption. In particular, research on multi-context systems [13] facilitates the integration of knowledge with heterogeneous logical representations such as envisioned in the sketched example scenario. Presumably, an information owner may employ not only structured or at least semi-structured information services but also ad-hoc facilities like email conversation led in natural language. To avoid the resulting opening of side-channels, we would need additional expertise, e.g., to convert "freely" expressed information into "more structural" one, as explored in the field of information extraction [21].

Guided by the two ideas sketched above and adapting the respective technologies, we come up with the architectural design described in Section 4. To actually build a manageable system we suggest to employ the idea of modern *software development and maintenance*, in particular to cope with specialization and parametrization as well as automatic optimization and administration, as further exemplified in Section 5. Additional insight can be provided by the field of multiagent systems [23], in particular for the integration of agent systems with other technologies, including standards for communication protocols [16].

We willingly accept to exercise no direct control at all on the cooperating agents and, accordingly, we only indirectly deal with options of collusion among those agents. These evident shortcomings will remain unsolved.

We intend to complement other approaches to overcome their restrictions and shortcomings sketched as follows. *Access control* and *encryption* applied to single data items are helpful but in general not sufficient to protect against an intelligent agent that might combine data received and already available before and intelligently *infer* consequences. *Usage control* is conceptually mandatory in general, but requires to implant trusted components into the interacting devices, see, e.g., [20]. Cryptographic *multiparty computations* are powerful means for protecting numerically encoded information but in most cases are rather costly and not applicable for logic-oriented information, see, e.g., [17].

3   Summary of Controlled Interaction Execution

Specifically realizing *security automata*, see, e.g., [19], for a logic-oriented view on information services, our own approach of inference-usability confinement by *Controlled Interaction Execution*, CIE, has been summarized in [3,4]. This approach originated from a seminal proposal of Sicherman/de Jonge/van de Riet [22] in 1983, which later has been resumed and extended by Biskup/Bonatti, e.g., [5,7] and then has further been elaborated by Biskup et al, e.g., [11,8,12,2,9,10]. CIE has been proved to be in accordance with fundamental notions of secrecy, as unified by Halpern/O'Neill [15], while adding dedicated logic-oriented enforcement mechanisms.

We briefly summarize the main characteristics of CIE concepts:

- *cooperativeness:*
  - no intervention whatsoever at other agents (seen as potential attackers),
  - only self-confinement when releasing own information,
  - confidentiality requirements as exceptions from permissions to share data;
- *logic-orientation:*
  - information represented by (sets of) sentences of a suitable logic, coming along with formal semantics of formulas, to precisely capture notions of knowledge and belief,
  - information acquired either explicitly/directly from communication data or implicitly/indirectly inferred by intelligent reasoning,
  - focus on possibilistic secrecy of a sentence to be kept confidential, roughly meaning, belief in the possibility of the sentence being not valid from an attacker's point of view;
- *statefulness:*
  - rich supported functionality for interaction to share data, including query answering, update processing with refreshments, belief revision, and data publishing, potentially as well as related services like, e.g., data mining,
  - unlimited interaction sequences,
  - keeping track of interaction history and thus state-based reactions;
- *modelling of "attacking" agent:*
  - agent-specific assumptions and agent-specific policy,
  - several approaches to attacker-specific reactions on potentially harmful requests, namely refusal, weakening, lying and combinations thereof,
  - simulating an attacking agent's postulated reasoning about candidates for reactions on an attacker's request to determine potential harmfulness, before deciding how to actually react;
- *formal assurances:*
  - formally proved compliance with confidentiality requirements under specified assumptions and policies.

In the remainder of this note, we elaborate our vision of a defending agent of the wanted kind, based on the broad theoretical work on CIE and preliminary experiences with an ongoing prototype implementation for a less ambitious situation, only requiring a frontend to a single relational DBMS.

4   Architectural Design

As a starting point, we assume that there are one or more existing functional components for information services – like a local DBMS for a private company, XML documents for personal and family data, and electronic health records. Figure 2 then shows the overall design of an agent implementing CIE. That agent should uniformly shield the functional components as a common control frontend to confine the outgoing flow of information to each of the cooperating agents according to the pertinent agent-specific policy and assumptions.

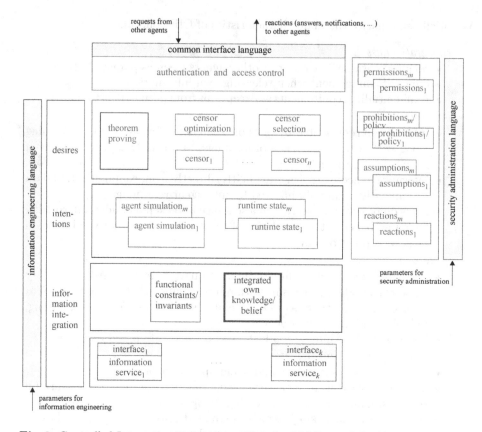

Fig. 2. Controlled Interaction Execution uniformly shielding a defending agent's integrated own knowledge or belief by means of, e.g., n available censors regarding, e.g., k existing information services offered to, e.g., m currently cooperating agents

As a prerequisite, we need a technology of *information system integration*, e.g., [14,13], to treat the existing functional components in a uniform manner:

- embedding each individual interface into a *common interface*;
- forming the integrated own *knowledge/belief* – whether explicitly or implicitly – , part of which is the target of the policy for prohibitions;
- evaluating an incoming request expressed in the common interface language in terms of evaluations of subrequests to pertinent *information services*.

Though highly demanding in its own right, it is indispensable to form a *unified own knowledge/belief* to which all security measures should refer, in order to achieve consistent enforcement of the owner's interests in confidentiality, independently of the information services involved and the interactions requested.

Each of the cooperating agents is specifically treated, as discretionarily specified by the defending agent's owner acting as security officer by means of context-specific parameters for the following components: *permissions* granted to the agent, the *confidentiality policy* as *prohibitions, assumptions* about the agent, and the kind of *reactions* in case of violating requests. Given these specifications, the *runtime state* and the *simulation* of the cooperating agent are initialized, and an initial *censor* is selected to control the reactions to that agent.

The runtime state represents the cooperating agent, in particular to capture original parameters, possibly their modifications, and past and current requests. Based on the runtime state, over the time, the component of *censor selection* determines the current *censor* for the agent. Together with the reactions shown to the agent, the runtime state also determines the *simulation* of the agent: basically, this simulation serves to model the cooperating agent's current belief about the defending agent's integrated own knowledge/belief, aiming to ensure invariantly that the former never contains any part of the latter that the policy requires to keep confidential.

In fact, the simulation of a cooperating agent('s belief of the defender's own knowledge/belief) is most crucial for an effective self-control as favored by CIE: though an agent is treated as cooperating in principle, it is also seen as a "curious attacker" and, as such, it cannot be supposed to frankly tell its belief about the defender's knowledge/belief. So, agent modelling by the control component is the only alternative, and obviously it can only be based on two features: the assumptions as specified and the requests and reactions having been occurred. Since the control component has observed the requests and even generated the reactions, it is completely certain about this feature; however, the assumptions are inherently uncertain. Accordingly, the parameters for the assumptions and their representation within the runtime state should be expressive enough to capture all relevant aspects of many and diverse situations; moreover, expressiveness should come along with best achievable orthogonality of dimensions to enable automatic translation into a runtime state and later uniform processing for censor selection and even *censor optimization* as well as agent simulation.

Besides the always postulated attacker's *system awareness* according to "no security by obscurity", i.e., knowledge of both the functional components and the confining frontend, the following parameter dimensions are most important:

- the attacker's *configuration awareness*: knowledge of specific declarations (interaction language, policy (security constraints), functional constraints);
- the attacker's applicable *common knowledge*: knowledge regarding the application, in particular schema declarations;
- the attacker's *specific knowledge*: knowledge resulting from other sources etc.;
- the attacker's *specific reasoning*: "procedural knowledge" to form beliefs etc.;
- the attacker's *guess* of the defender's hidden parameters (kind of defender's functional reasoning etc.).

Notably, in principle not only the defending agent to be implemented is uncertain about an attacking agent's belief, but also the attacking agent is uncertain how it is simulated by the defender, and thus there is mutually recursive uncertainty to be suitably resolved. Indeed, each particular censor working with a specific simulation has to be justified by a convincing postulate on the *coincidence* of the actual attacker with the defender's simulation.

5   Uniformity for Specific Engineering Tasks

Within the design, we will have to treat many specific engineering tasks. As exemplified in the following, for each task we envision to achieve *uniformity* across the anticipated variety of situations: conceptually, the situations are captured by a powerful *abstraction* which then, by an implementation, is strictly *encapsulated* into a single module; further, the selection of a specific situation is enabled by a powerful *parameterization* and embodied by a *specialization* of that module.

Existing information services may considerably vary in the *underlying logic*, which, basically, defines the specific semantics. For example, an information service might be based on a *completeness* assumption – enabling reasoning about a "closed-world" [5,6,12]–, faced with partial or principle *incompleteness* – restricting or even disabling reasoning about "negative information" [11,9] –, or equipped with a concept of *preferences* – introducing essential differences between certain knowledge and uncertain belief and requiring non-monotonic reasoning for revisions and updates [10]. Given the pertinent parameters for the information services to be integrated, the totality of information available to the owner is abstracted into the integrated own knowledge/belief, and querying and anticipated manipulations, respectively, are encapsulated by specific modules, which have appropriate specializations for the potentially occurring variations.

Under each of these variations, a censor basically has to evaluate possible reactions on a request regarding *harmfulness* of adding the supplied information to the requester's current belief. This belief is abstracted as the pertinent agent simulation, and the needed evaluations are encapsulated by a module for determining harmfulness. Again, this module has appropriate specializations, which might employ the incorporated theorem prover to reduce the problem of harmfulness to a suitable entailment problem in the pertinent logic.

The abstraction of an agent simulation permits a large range of actual implementations. In a simple dynamic case, there is just a *logfile* containing sentences reflecting the a priori knowledge and the reactions provided so far [5,6,11]. For optimization, instead of keeping a logfile an *adapted version* of the original confidentiality policy might be maintained [2]. In the more advanced context of non-monotonic belief management, the agent simulation comprises both an *approximation* of the own knowledge/belief and *skeptical reasoning* regarding the aspects left only approximated [10]. For the static case of data publishing, a possibly distorted, "inference-free" *alternative view* on the own knowledge/belief is generated beforehand and later employed like an agent simulation, for which the associated censor does not need to perform any further dynamic control [12,9].

6   Experiences, Further Issues, and Concluding Remarks

We are implementing a *CIE-prototype* [3,4] for a simplified scenario, only shielding a single *relational DBMS*, Oracle, for somehow restricted interactions. This prototype provides a uniform treatment of all included interactions for various kinds of cooperating agents and the permissions, prohibitions (policies), assumptions and reactions specified for them. For each such a situation, the actual control is established by a dedicated specialization of a general censor component.

Successful CIE operation needs powerful tools and facilities for the *administration* of agent-specific parameters, and (semi-)automatic *optimization*. In particular, the initial selection of an appropriate censor instance followed by repeated reconsiderations appear to be crucial. As expected by theoretical insight, computational complexity and scalability remain a major issue. Accordingly, identification of parameters leading to feasible cases and their automatic recognition as part of optimization are further important topics. Our experiences also suggest to sometimes refrain from the principle of minimal distortion, but instead to look for efficiently computable overestimations of an agent's simulation.

A main concern is to achieve *robustness* of the defending agent's attempt to simulate the postulated behavior and reasoning of another agent seen as attacker: what assurances regarding preservation of confidentiality can the defending agent get if the attacker's situation actually differs from the defender's simulation?

There are several further issues whose solutions might have an impact on both the architectural design and specific tasks. Basically, in each case we would have to decide whether to include additional control components or only refined parameters for the already existing components, the latter found to be extremely worthwhile so far and hoped to be extendable. In the following, we briefly list selected issues: *collusion* among cooperating agents has to be made useless by treating a group of agents like a single agent; possibilistic confidentiality might be strengthened, for example to k-*confidentiality* demanding a stronger "negative belief", by ensuring the existence of at least $k > 0$ essentially different counterexamples, or to *probabilistic secrecy* or related notions; we might also weaken our current notion into a sort of *complexity-theoretic secrecy*; seen from a broader perspective, the *desires* and *intentions* of a BDI-like agent [23] might influence the agent-specific parameters, as do *normative concepts* [1]; as usual, production of *reliable software* and its appropriate *installation* are mandatory, leaving no options to circumvent the censoring or to exploit side-channels.

Concluding, we advocate the engineering of inference control as a *frontend* to existing functional components as a promising step to our vision, emphasizing a *uniform treatment* of various situations by means of strict *encapsulation* of powerful *abstractions*, expressive *parameterization*, and the concept of *specialization*.

References

1. Andrighetto, G., Governatori, G., Noriega, P., van der Torre, L.W.N. (eds.): Normative Multi-Agent Systems. Dagstuhl Follow-Ups, vol. 4. Schloss Dagstuhl – Leibniz-Zentrum für Informatik (2013)
2. Biskup, J.: Dynamic policy adaption for inference control of queries to a propositional information system. Journal of Computer Security 20, 509–546 (2012)
3. Biskup, J.: Inference-usability confinement by maintaining inference-proof views of an information system. International Journal of Computational Science and Engineering 7(1), 17–37 (2012)

4. Biskup, J.: Logic-oriented confidentiality policies for controlled interaction execution. In: Madaan, A., Kikuchi, S., Bhalla, S. (eds.) DNIS 2013. LNCS, vol. 7813, pp. 1–22. Springer, Heidelberg (2013)
5. Biskup, J., Bonatti, P.A.: Controlled query evaluation for enforcing confidentiality in complete information systems. Int. J. Inf. Sec. 3(1), 14–27 (2004)
6. Biskup, J., Bonatti, P.A.: Controlled query evaluation for known policies by combining lying and refusal. Ann. Math. Artif. Intell. 40(1-2), 37–62 (2004)
7. Biskup, J., Bonatti, P.A.: Controlled query evaluation with open queries for a decidable relational submodel. Ann. Math. Artif. Intell. 50(1-2), 39–77 (2007)
8. Biskup, J., Gogolin, C., Seiler, J., Weibert, T.: Inference-proof view update transactions with forwarded refreshments. Journal of Computer Security 19, 487–529 (2011)
9. Biskup, J., Li, L.: On inference-proof view processing of XML documents. IEEE Trans. Dependable Sec. Comput. 10(2), 99–113 (2013)
10. Biskup, J., Tadros, C.: Preserving confidentiality while reacting on iterated queries and belief revisions. Ann. Math. Artif. Intell. (2013), doi:10.1007/s10472-013-9374-6
11. Biskup, J., Weibert, T.: Keeping secrets in incomplete databases. Int. J. Inf. Sec. 7(3), 199–217 (2008)
12. Biskup, J., Wiese, L.: A sound and complete model-generation procedure for consistent and confidentiality-preserving databases. Theoretical Computer Science 412, 4044–4072 (2011)
13. Brewka, G.: Multi-context systems: Specifying the interaction of knowledge bases declaratively. In: Krötzsch, M., Straccia, U. (eds.) RR 2012. LNCS, vol. 7497, pp. 1–4. Springer, Heidelberg (2012)
14. Calvanese, D., Giacomo, G.D., Lenzerini, M., Rosati, R.: View-based query answering in description logics: Semantics and complexity. J. Comput. Syst. Sci. 78(1), 26–46 (2012)
15. Halpern, J.Y., O'Neill, K.R.: Secrecy in multiagent systems. ACM Trans. Inf. Syst. Secur. 12(1), 5.1–5.47 (2008)
16. Huget, M.-P., Poslad, S.: The Foundation of Intelligent Physical Agents, http://www.fipa.org
17. Kolesnikov, V., Sadeghi, A.-R., Schneider, T.: A systematic approach to practically efficient general two-party secure function evaluation protocols and their modular design. Journal of Computer Security 21(2), 283–315 (2013)
18. Kott, A., McEneaney, W.M. (eds.): Adversarial Reasoning: Computational Approaches to Reading the Opponent's Mind. Chapman & Hall/CRC, Boca Raton (2007)
19. Ligatti, J., Bauer, L., Walker, D.: Run-time enforcement of nonsafety policies. ACM Trans. Inf. Syst. Secur. 12(3) (2009)
20. Pretschner, A., Hilty, M., Basin, D.A.: Distributed usage control. Commun. ACM 49(9), 39–44 (2006)
21. Sarawagi, S.: Information extraction. Foundations and Trends in Databases 1(3), 261–377 (2008)
22. Sicherman, G.L., de Jonge, W., van de Riet, R.P.: Answering queries without revealing secrets. ACM Trans. Database Syst. 8(1), 41–59 (1983)
23. Wooldridge, M.J.: An Introduction to MultiAgent Systems, 2nd edn. Wiley, Hoboken (2009)

Automated Formal Verification
of Application-specific Security Properties

Piergiuseppe Bettassa Copet and Riccardo Sisto

Dipartimento di Automatica e Informatica
Politecnico di Torino, Italy
{piergiuseppe.bettassa,riccardo.sisto}@polito.it

Abstract. In the past, formal verification of security properties of distributed applications has been mostly targeted to security protocols and generic security properties, like confidentiality and authenticity.

At ESSOS 2010, Moebius et. al. presented an approach for developing Java applications with formally verified application-specific security properties. That method, however, is based on an interactive theorem prover, which is not automatic and requires considerable expertise. This paper shows that a similar result can be achieved in a fully automated way, using a different model-driven approach and state-of-the-art automated verification tools. The proposed method splits the verification problem into two independent sub-problems using compositional verification techniques and exploits one tool for analyzing the security protocol under active attackers and another tool for verifying the application logic. The same case study that was verified in the previous work is used here in order to show how the new approach works.

1 Introduction

Formal verification of security properties of distributed applications is attracting researchers' attention, especially in recent years, because of the increasing diffusion of applications with important security requirements.

Distributed applications with security requirements generally use cryptographic protocols to communicate over insecure channels. Sometimes the security protocol and the application logic are totally independent: the protocol provides virtual communication channels with standard security properties (mutual authentication, confidentiality, data integrity) and the application is developed in a nearly security-unaware way, security being provided just by application insulation, which is guaranteed by the fact that the application communicates only over secure channels. In other cases, however, protocol and application logic are less independent. For example, custom protocols can be used in order to guarantee application-specific properties, and the application logic may interact more strictly with the protocol in order to achieve the desired security properties. Of course, using an independent security layer realized by standard protocols (for instance TLS) is preferred when possible, because of its simplicity and reliability. However, this is not always possible or convenient, for example because the devices involved do not have

J. Jürjens, F. Piessens, and N. Bielova (Eds.): ESSoS 2014, LNCS 8364, pp. 45–59, 2014.

enough hardware resources or do not have standard connectivity to the Internet, but only limited ad-hoc connectivity.

The techniques and tools for automated formal verification developed so far are mostly targeted to either the analysis of security protocols or the analysis of application code. On the one hand, some tools [1] can formally verify standard security properties of cryptographic protocols under the presence of active attackers. However, these tools can analyze only the bare protocol (message exchanges and related checks) while they are not adequate to also model and analyze the application logic that interacts with the protocol, which can be made of complex programs, without particular constraints. Moreover, generally these tools cannot deal with application-specific security properties. On the other hand, tools for the automated formal verification of arbitrary application source code are available (e.g. software model checkers [2]). In theory these tools even allow to consider active attackers in the system, but a model of those attackers must be supplied by the user and the inclusion of active attackers makes verification very complex.

Actually, the main obstacle to extending existing verification techniques to analyze both protocol and application logic together in the face of active attackers is mainly practical, and is related to the limited scalability of these verification techniques. In fact, the problem of cryptographic protocol verification is itself challenging despite the simplicity of such protocols.

A case study of formal verification of application-specific security properties (i.e. the truth of a predicate involving some variables of the application), taking into account both the protocol and the application logic together, appeared recently in literature [3]. In this case study the application is developed with a model-driven approach and the model is used to generate a formal specification, which afterwards can be verified by an interactive theorem prover. An important limitation of this approach is that it is based on interactive theorem proving, which is not automatic, is very time consuming, and requires a lot of expertise. Moreover, if the application is flawed, interactive theorem proving does not provide counter examples, which can make error diagnosis and correction very difficult.

In this paper we show that a simpler approach can be used to achieve a similar result. In fact, the proposed method is based on verification techniques that are automated, simpler to use, and that can also provide counter examples when the properties to be verified do not hold.

The main idea is to combine two already existing and well-known automated formal verification techniques, *theorem proving* for cryptographic protocol verification and *model checking* for source code verification, according to the principles of assume-guarantee compositional verification. This approach brings, in addition to the above mentioned advantages, better scalability, due to the splitting of the verification problem into simpler sub-problems. The work proposed in this paper, as well as combining the two mentioned verification techniques, also aims at automating the entire process of implementing and verifying distributed applications. To our knowledge, at present there are no other proposals with the same characteristics in literature.

The proposed development approach is based on the principles of model driven design: it starts by defining a high-level formal model of the communication protocol, where also the expected security properties of the protocol are formally specified. An automated formal verification of those properties is performed by the protocol verifier ProVerif [4], and the Java implementation of the protocol is automatically generated by the model driven development framework JavaSPI [5], which guarantees the preservation of the intended security properties. The resulting protocol implementation must then be integrated within the application logic (client and server), which can be developed in any way (hand written or developed using other code generation techniques). Then, the application-specific properties are formulated and verified on the application logic using a Java source code verifier, such as Java Pathfinder (JPF)[2], but taking the results of the protocol formal verification into account. This is achieved by replacing the code that implements the protocol with a stub that enforces the properties already verified on the protocol model. If compared to a separate and independent use of the theorem prover and the model checker, the main advantage of the methodology proposed here is the reduction of verification complexity, made possible by leveraging compositional verification in the assume-guarantee reasoning style.

The whole process evolves in a largely automated workflow, which reduces the probability of introducing errors significantly, and enables quick error diagnosis (both the tools used for formal verification can provide counter examples, i.e. the execution traces that violate the intended properties).

The remainder of the paper is organized as follows. Section 2 discusses related work and Section 3 gives some background about the tools that are exploited in this work (ProVerif, JavaSPI and Java Pathfinder). Then, Section 4 explains some new features that have been added to JavaSPI in order to support the approach proposed in this paper and Section 5 introduces the case study example and describes how it was developed and verified. Finally, Section 6 concludes.

2   Related Work

In the last decades many automated techniques have been developed for the formal analysis of security protocols, as recently surveyed in Patel et al. [1]. These techniques analyze high-level abstract models, in order to prove the correctness of the protocol logic. More recently, some researchers have started working on techniques that bring automated formal proofs closer to real implementations of security protocols [6]. Among these are the model-driven development approaches, like the one exploited in this paper [5].

All the above mentioned techniques are focused on security protocols rather than on whole applications, and address the generic security properties enforced by such protocols (e.g. authentication, secrecy and integrity), rather than the application-specific security properties.

Some papers have addressed the formal verification of security protocols for specific applications, such as for example electronic commerce, with their related application-specific properties. For example, Bella et al. [7] presented the formal verification of some application-specific properties of the suite of protocols "Electronic Secure Transaction", used for e-commerce. However, this work is substantially different from the one presented here because verification is not automatic (being based on the interactive theorem prover Isabelle [8] which requires human assistance), and what is formally verified is only an abstract model of the application rather than its final implementation.

Besides the work by Moebius et al. [3] that was already mentioned in the introduction, and a related publication [9] that presents exactly the same methodology but applied to a service-oriented application, some other papers have addressed the problem of developing distributed applications with formally verified security properties. A recent paper [10] extends the previous approach by integrating the AVANTSSAR [11] model checker into SecureMDD. As a result, it is possible to automatically generate a formal specification for the model checker from a UML model. However, only some application-specific properties can be verified using AVANTSSAR. For example, differently from the work presented here, which enables the verification of arbitrary properties, it is not possible to compare numeric values inside the model checker.

Jürjens [12] proposed a UML-based technique for the specification of distributed applications and automated formal verification of application-specific security properties. The technique was applied to the Common Electronic Purse Specifications regarding payment via smart-card. One of the properties that were verified is, for example, that the amount of money in the system is every time the same, that is the total sum of budgets of smart-card holders is always equal to the sum of the earnings of all merchants. However, this technique provides formal verification of UML models only, whereas a formal link with the application implementation is missing. Moreover, differently from our approach, verification is performed in a single step on the whole model, without using compositional verification.

Gunawan et al. [13] proposed a method to integrate some standard security mechanisms (for protecting information transfer) into distributed applications automatically. The paper includes a proof that the security mechanisms are integrated into the application so as to fulfill some generic properties. However this approach does not target the verification of application-specific properties.

The idea of using compositional verification to formally verify application-specific security properties of distributed applications already appeared in Gunawan and Herrmann [14]. In that work, however, formal verification is done by a general-purpose model checker, without considering active network attackers and the properties of cryptographic operations.

3   Background

3.1   ProVerif

ProVerif [4] is an automated theorem prover for cryptographic protocols. In ProVerif, the protocol and the attacker are modeled according to the Dolev-Yao

[15] symbolic approach, which substantially means representing data and cryptographic operations symbolically and assuming the attacker has complete control over public communication channels, thus being able to read, delete, and modify messages in transit or forge new messages using the knowledge the attacker has achieved so far. The symbolic representation of data and cryptography entails that cryptography is assumed to be ideal. For example, an encrypted message can be decrypted only if the correct decryption key is known. Differently from model checkers, ProVerif can model and analyze an unbounded number of concurrent sessions of the protocol, thus providing results that hold for any number of parallel sessions. However, like model checkers, ProVerif can reconstruct a possible attack trace when it detects a violation of the intended security properties. ProVerif may report false attacks, that is attacks that in reality are not possible, but at the same time if a security property is reported as satisfied then it is true in all cases, so it is necessary to analyze the results carefully when attacks are reported.

3.2   The JavaSPI Framework

JavaSPI [5] is a framework for modeling, formally verifying and implementing cryptographic protocols, according to the paradigm of model-driven development. Initially, the user defines an abstract formal model of the protocol according to the Dolev-Yao modeling approach. This model, being abstract, does not include implementation details such as, for example, hash algorithms and length of cryptographic keys. This model can be formally verified by ProVerif in order to check that it satisfies some security properties. These properties are generally expressed either as secrecy requirements (the attacker must not be able to know some data) or as correspondence requirements referred to events specified in the abstract model. The latter requirements can be used to express authentication or data integrity properties; for example an authentication requirement could be expressed as $terminate(A, B) \Rightarrow start(B, A)$, which means that each time actor A terminates a session of the protocol apparently with B (i.e. event $terminate(A, B)$ occurs), B has previously started a session of the protocol with A (i.e. event $start(B, A)$ has occurred).

When the user is satisfied with the model and confident about its logical correctness, the missing implementation details can be specified and a Java implementation of the protocol can be automatically generated. JavaSPI is very similar to Spi2Java [16], the main difference being the modeling language: while with Spi2Java a protocol is modeled directly in the formal specification language spi-calculus, JavaSPI lets the user develop the protocol model in the form of a Java application, written with some restrictions on the Java language and making use of a custom library (JavaSpiSim), which offers the same expressiveness as the spi calculus language. In fact, a formal specification of the protocol compatible with ProVerif can be generated automatically from the Java code. Using Java as the modeling language facilitates users who are familiar with object oriented programming and Java. Moreover, this approach lets the user simulate the execution logic of the protocol by means of a normal Java debugger.

Figure 1 shows an excerpt of an abstract model written with JavaSPI. Each model is composed of a number of processes, each one specified by a Java class that extends the spiProcess library class. The behavior of a process is specified by defining the doRun method, which takes as arguments objects belonging to classes of the JavaSpiSim library. These classes represent the data types admitted in a security protocol model and include methods for performing common operations, such as for example encrypting or decrypting data or sending or receiving data on channels. The occurrence of an event is specified by calling the event method which can have any number of arguments (e.g. event("start",A,B) generates event $start(A, B)$.

The implementation details that are necessary for generating the final implementation code are specified as Java annotations added to the abstract model. JavaSPI shares with Spi2Java the same code generation mechanism, which has been proved to preserve a large class of security properties [17]. This means that if a security property has been proved to hold on the formal model, then that property holds on the automatically generated Java implementation too.

3.3   Java Pathfinder

Java Pathfinder [2] (JPF) is a software model checking tool for the Java language. Java Pathfinder can directly analyze the bytecode of Java multithreaded applications, checking the truth of assertions or LTL formulas. Java Pathfinder consists of a particular Java Virtual Machine (JVM) which executes the bytecode by exploring all possible execution paths (when nondeterministic choices are possible in the execution, each one of them is explored by backtracking execution).

JPF includes several optimizations that automatically reduce the number of states to be visited (avoiding those whose inspection is redundant) and thus the complexity of the analysis.

4   The Extended JavaSPI

To achieve the final goal of this work the JavaSPI framework has been extended in order to enable increased interaction between the generated protocol code and the application that uses the protocol. With the original JavaSPI, only a simple interaction mechanism was possible, where the application starts a protocol session, passing input arguments, and, upon termination of the protocol session, the application gets the outputs. With the extended JavaSPI version, the application can be called back by the protocol code when some events defined in the model occur. In this way, the application can receive outputs from the protocol at intermediate stages of a protocol session. The @EventsInterface annotation enables this new mechanism. When the annotation is present, the code generator generates a Java interface that contains the methods associated with the events generated by the process and has the name specified in the annotation. When a session of the protocol is started by the application, a callback object that

```
public class p_Card extends spiProcess {
...
  @EventsInterface("p_Card_Interface")
  public void doRun(Channel cTermCard, Nonce passphrase,
          Identifier LOAD, Identifier PAY, Identifier TERMAUTH,
          Identifier RESAUTH, Identifier TERMLOAD, Identifier TERMPAY,
          Identifier RESPAY) throws SpiWrapperSimException{

   Message xIn = cTermCard.receive(Message.class);

   if(xIn.equals(TERMAUTH)){
     Nonce challenge = new Nonce();
     Pair<Identifier,Nonce> _w0 = new Pair<Identifier, Nonce>(RESAUTH,challenge);
     cTermCard.send(_w0);

     Pair<Message,Hashing> _p0 = cTermCard.receive(Pair.class);
     Pair<Identifier,Integer> xTermLoad_xValue = (Pair<Identifier, Integer>) _p0.getLeft();
     Hashing xHash = _p0.getRight();
     Identifier xTermLoad = xTermLoad_xValue.getLeft();
     Integer xValue = xTermLoad_xValue.getRight();

     if(xTermLoad.equals(TERMLOAD)){
       Pair<Identifier,Nonce> _w1 = new Pair<Identifier, Nonce> (LOAD,passphrase);
       Pair<Message,Nonce> _w2 = new Pair<Message, Nonce>(_w1,challenge);
       Pair<Message,Integer> _w3 = new Pair<Message, Integer>(_w2,xValue);
       Hashing h = new Hashing(_w3);

       if(h.equals(xHash)){
         event("addToBalance",xValue);
...
```

Fig. 1. Excerpt of a sample model code

implements the generated interface must be passed as argument. This extension does not affect the validity of the ProVerif model that is generated from JavaSPI, because the methods called on event occurrence cannot alter the protocol behavior as modeled by ProVerif. As detailed in Section 5.6, when performing the verification of the application code, the protocol code is substituted by stubs that enforce exactly the event orderings that are made possible by the protocol.

5 The Case Study Application Development

The case study is the development of a smart-card based application that implements a sort of electronic purse. The application lets the user load credit onto the smart card and use the loaded credit to get some services. In Moebius et al. [3], a copy service offered by a University Campus to students is considered, but which specific service is offered by the application is not relevant. In the description of the case study, we stick to the copy service example.

The users of the application are some customers and a manager. Each customer owns a smart-card where Java code can run, on which credit can be loaded. The manager provides a set of terminals where customers can go with their smart-card in order to buy or spend credit. The current balance of credit is stored on the smart card and is updated at each operation performed. For simplicity, the example considers one unit of credit corresponding to one copy.

Finally, all terminals and all smart-cards store internally the same secret key, shared by all trusted and original components. The secret keys are assumed to be not accessible, both in the smart-cards and in the terminals (the smart-card is assumed to be tamper-proof while the terminal is assumed to be secured so that only the manager can access its internals for maintenance).

The security goal that is considered in this case study is "the manager does not lose money", that is the total amount of issued copies does not exceed the total credit bought previously by all users during their loading operations on their smart-cards. This property must be satisfied even in the presence of potential active attackers who may intercept/alter/delete messages transmitted between the actors (smart-cards and terminals), or create new ones, following the definition of attackers of the Dolev-Yao model.

5.1   The Development Workflow

The key idea of the proposed development approach (depicted in Figure 2) is to divide the application into two distinct parts, to be developed and verified separately: the protocol, and the application logic.

The protocol is developed according to the JavaSPI model-driven methodology. It includes all communication activities and must satisfy some security properties, specified by the developer.

The application logic can be developed in any way, but it must properly interact with the protocol, by starting protocol sessions and reacting to events.

The verification process is compositional. The security properties of the protocol are verified on the abstract protocol model using ProVerif and assuming a generic scenario with an unbounded number of parallel protocol sessions. The same properties are guaranteed to hold on the Java code that implements the protocol by the code generation algorithm. Application-specific security properties are specified and formally verified using an automated formal verification tool capable of analyzing Java code directly (Java Pathfinder in our case). When performing this verification step, it is possible to avoid the explicit modeling of the protocol part, by substituting it with a stub that describes the security properties proved by ProVerif. The stub can be automatically generated from the protocol properties.

The rest of this section details the various steps with reference to the case study.

5.2   Developing the JavaSPI Abstract Protocol Model

The protocol designed for this application is based on challenge interactions. Figure 3 shows the interaction between a terminal and a card during the load operation.

Once a card is plugged into the terminal, the user can enter money into the terminal, which triggers the start of the load operation. This operation then proceeds as shown in Figure 3, where *value* is the amount of credit to be loaded. The terminal starts the operation generating the *addToIssued(value)* event and

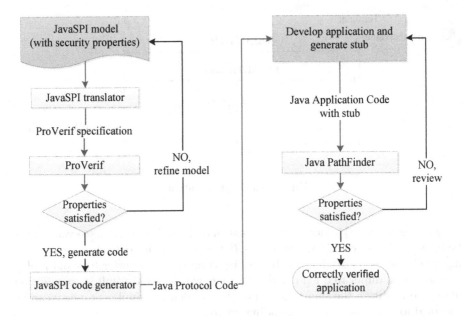

Fig. 2. Workflow of the verification process

sending the card the TERMAUTH message. The card responds issuing the challenge message, composed of the RESAUTH tag and a nonce (a randomly generated number). The terminal responds to the challenge by sending the last message, which includes the TERMLOAD tag followed by the value to be loaded and a hash value, computed on a 4-tuple that includes the shared secret key, the nonce and the value. Finally, the card re-computes the hash value using its own copy of the secret key and nonce and the value received in the message, and if the result matches the received hash value it concludes successfully the operation, by generating the *addToBalance(value)* event.

The two events will correspond to operations in the application logic that record, respectively, the amount of money earned and the amount of credit spent.

The JavaSPI specification of the card behavior during the load operation is the code excerpt shown in Figure 1.

The JavaSPI model can be simulated in order to check that it behaves as expected.

This security protocol is expected to satisfy two main security properties. The first one is that the secret shared by all the original components cannot be known by an attacker, who has access to the communication channel between the terminal and the card. The second one is the correspondence of the protocol events. For the load operation, each time some credit is actually loaded onto a smart-card (event *addToBalance(credit)*), the corresponding amount of money must have been previously entered into one terminal (event

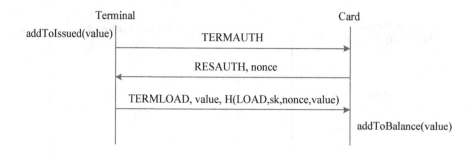

Fig. 3. The load operation

addToIssued(credit)). Moreover, the correspondence between these events must be injective, i.e. any *addToBalance(credit)* event must have its own corresponding *addToIssued(credit)* event. Injectivity is necessary in order to avoid replay attacks (i.e. a duplicated load credit message, which would result in an addition of unpaid credit on the smart-card, must be avoided). A similar property can be specified for the operation of spending credit.

Note that a tool like ProVerif cannot model integer arithmetic and precedence comparisons between integers. Hence, it does not allow to specify more complex properties, e.g. the ones related to the sum of credit loaded or spent, nor it allows to describe the application logic that processes the events and updates integer counters.

5.3   Formal Protocol Verification

The model generated in the previous step is automatically converted by JavaSPI into the input syntax accepted by ProVerif. The resulting code is ready to be formally analyzed, but first the information on the multiplicity of processes must be added, in order to indicate that there may be an unbounded number of instances of processes.

ProVerif succeeds in proving that the intended properties of the protocol hold on the model. ProVerif takes 15ms to complete the proof on a computer equipped with Intel Core2 Quad Q9450 running at 2.66GHz, 8 GB of DDR2 RAM and Ubuntu 12.04 64-bit operating system and ProVerif 1.86p13.

5.4   Protocol Code Generation

After having verified the model with ProVerif, the generation of the Java code that implements the protocol can take place, by means of the code generator provided by JavaSPI. The result is a set of Java packages, one for each process in the model, which implements the behavior defined in the model.

5.5   Application Logic Development

The generated protocol code must now be integrated with the application code that uses it. In our case study, the application code has been kept simple, but it includes all the fundamental aspects of the application that are necessary for its verification. More precisely, only the functionalities related to the management of the credit system have been implemented on the card software.

5.6   Checking the Application Code

The last step of the workflow is the verification of the application-specific properties using Java Pathfinder. As already anticipated, in order to reduce the complexity of this verification task, the protocol code generated by JavaSPI is replaced with a stub that just reproduces any possible behavior of the protocol sessions, as seen by the application, without really executing the protocol. Of course, the behavior of the stub must be constrained so as to satisfy the security properties that have already been verified by ProVerif. In principle, this constraint can be enforced in one of two different ways: either the constraint is enforced when generating the stub, or the stub is generated without any constraint but the application-specific security property P to be verified is rewritten in the following form

$$C \Rightarrow P$$

where C is the constraint (i.e. the property verified by ProVerif). This second approach is more difficult, because of the difficulty of expressing C. Then, the first approach (generation of a stub that incorporates the constraints coming from the properties verified by ProVerif has been selected for our case study).

As the application processes interact with each other only through the protocol, having replaced the protocol implementation with the stub makes it possible to avoid considering the behavior of active attackers any more during the verification of the application code. In fact, the behavior of potential active attackers has already been considered when analyzing the protocol by ProVerif, and it is already incorporated in the stub behavior itself.

Based on the architecture of the developed application, the only possible interactions between the protocol and the application logic are those that occur at the start and at the end of each session, as well as at the occurrence of one of the intermediate events described in the model. For this reason, it is enough for the stub to include the statements corresponding to these interaction points. All the other statements that make up the protocol implementation can be safely omitted.

The stub can be built by creating multiple Java threads, each one playing the behavior of a single actor in a single protocol session. In order to include the constraints deriving from the security properties verified by ProVerif, it is enough to synchronize these threads in such a way that the security properties proved for the protocol are enforced.

In our case study, the stub includes threads that play the role of the terminal and threads that play the role of the card. The threads that play the role of the terminal learn the kind of operation and the amount of credit to be loaded or spent at their startup (this information is an input coming from the user when the application starts the session). Instead, the threads that play the card role are ready to perform either a load or a spend operation, which in the real protocol is selected by the first message received.

If we want to constrain the behavior of these threads so as to enforce the correspondence properties that have been verified by ProVerif, we have to synchronize the events of each card thread with the events of a corresponding terminal thread. More precisely, before performing a load or spend event, a card thread has to synchronize with a terminal thread that has just performed a corresponding event. This means that the terminal thread enters a synchronization state after having generated an event while a card thread enters a synchronization state before proceeding with a load or spend event.

Model checking does not allow to analyze systems with an unbounded number of states. For this reason, a necessary condition is that the number of parallel protocol sessions (i.e. the number of threads in the stub) is kept bounded. In our case study, this corresponds to having bounded numbers of users and terminals (as each user has one card, the number of cards equals the number of users), with the assumption that no more than one session at a time is possible on each card or on each terminal.

In addition to bounding the number of threads, as with any software model checking problem, abstractions in the application code may be necessary, in order to make the number of states finite and reasonably small.

In our case study, the application-specific property to be checked is given by the fact that in every instant (or for every state reached and analyzed by the model checker) the value of an integer field (named "balance", which represents the difference between the current paid copies and those issued) is always greater than or equal to zero. This property details the more general property "the manager does not lose money". Since the instantaneous value of the balance field depends on the field additions and subtractions performed by the application itself, it is not possible to introduce a layer of abstraction on it. Nevertheless, it is still affordable to run the model checker over a reasonable number of possible cases.

To check if it is satisfied there are two possible ways.

The first one is to use a plugin for Java Pathfinder that enables the verification of LTL formulas during the state exploration performed by JPF. The plugin [1] used in this case study is not maintained directly by the JPF development team and is subject to discontinuity of development over the years. Other plugins that support LTL verification are available. However, the one used in this case study was chosen because it supports the verification of class field values, and not only method calls sequences. In this case, the LTL property to verify is specified through the following annotation:

---

[1] Available at `https://bitbucket.org/petercipov/jpf-ltl`

```
@LTLSpec("[] ( it.polito.javaSPI.test.CSJPF.balance>=0)")
```

where `it.polito.javaSPI.test.CSJPF` is the class that includes the balance field. This formula simply means that the balance is always greater than or equal to zero.

The second way is to introduce assertions within the application code. In this case, since the example application requires that the "balance" is always non-negative, it is sufficient to place an "`assert balance >= 0`" at any point in the code where the value of the "balance" is set or modified. As this is a private field, it is very simple to identify the only places where it can be set or modified.

Results show that both methods work well for our case study. No violations of the specified properties are detected, thus proving, by exhaustive state exploration, that the properties hold on the application code. Furthermore, the method that uses assertions occupies less memory (RAM) and takes less time, compared to the LTL formula verification.

Verification with JPF was performed on a computer equipped with Intel i7-3770 CPU running at 3.40GHz and 11GiB of DDR3 RAM. The software components relied on an Ubuntu 13.04 32-bit operating system, Java HotSpot(TM) Server VM (Java version 1.7.0_21, build 23.21-b01, mixed mode).

The initial JavaSPI model is composed by 250 lines of Java code and annotations. The size of the ProVerif model is 150 lines, and the size of the protocol code is about 450 lines of Java code. Both are generated by the JavaSPI generator starting from the initial model. The final application requires about 200 additional lines of Java code.

Table 1 and Table 2 report the time and memory required for the verification of the case study example, in the cases of assertions and LTL formula respectively. With the computational resources specified above, in this case it has been possible to analyze a scenario with a maximum of 4 users and 4 terminals when the LTL formula verification is performed. Conversely, the verification of the *assert* conditions requires fewer resources, and can handle efficiently systems with up to 5 users and 5 terminals. It is important to note, however, that, in general, assertions are not always enough for expressing application-specific properties. Therefore, in other case studied the use of LTL formulas can be unavoidable.

Although it is not possible, with a model checker, to formally infer that the properties hold with any number of users and terminals, the results obtained with a small number of participants are sufficient to give reasonable confidence that this is true. In fact, if a distributed application is flawed, usually the error can be detected even with small numbers of parallel sessions.

Table 1. Java Pathfinder verification time and memory consumption using the **assert** construct

	Users and terminals				
	1	2	3	4	5
Time	<1s	1s	7s	2m 40s	42m 21s
Memory	61MB	79MB	145MB	275MB	697MB

Table 2. Java Pathfinder verification time and memory consumption of the LTL formula

	Users and terminals			
	1	2	3	4
Time	1s	30s	30m 18s	44h 24m 59s
Memory	61MB	290MB	467MB	952MB

As mentioned above, the characteristics of the application itself have a significant effect on the complexity of model checking, so performance can be very different depending on the application under test.

6   Conclusions

In this paper it has been shown how a distributed application with application-specific security requirements can be developed using a model-driven approach that finally yields a formally verified Java implementation. The formal verification of the security properties takes into account active attackers and is entirely automated. The most critical part of the code, i.e. the implementation of the security protocol, is generated automatically from an abstract model with the guarantee of security property preservation. Moreover, the model is written in Java, instead of using domain-specific formal languages. The adoption of a compositional verification approach splits verification into two separate simpler tasks, which potentially leads to the possibility to handle larger applications.

Up to our knowledge, no other approach was previously proposed with all these features together. Compared to the approach presented in [3], which developed the same case study, our approach has the advantage of being fully automated. Even if model checking does not allow us to get a result that holds for any number of users and terminals, the result gives anyway good security assurance and can be obtained using only automated tools and without requiring excessive expertise.

The results obtained are encouraging because they confirm that it is possible to develop distributed applications with formally verified application-specific security properties using only automated tools.

One drawback that we found is the high quantity of resources that the model checking with JPF requires, in terms of memory and time. This is partially due to the kind of verification that interprets the bytecode of the real Java application. Using other verification tools for Java may improve the performance. Future works will address the verification of generic security properties in the final application code, for example guarantee that a the value of a field added manually remains confidential in the final application.

References

1. Patel, R., Borisaniya, B., Patel, A., Patel, D., Rajarajan, M., Zisman, A.: Comparative analysis of formal model checking tools for security protocol verification. In: Meghanathan, N., Boumerdassi, S., Chaki, N., Nagamalai, D. (eds.) CNSA 2010. CCIS, vol. 89, pp. 152–163. Springer, Heidelberg (2010)

2. Visser, W., Havelund, K., Brat, G., Park, S., Lerda, F.: Model checking programs. Automated Software Engg. 10(2), 203–232 (2003)
3. Moebius, N., Stenzel, K., Reif, W.: Formal verification of application-specific security properties in a model-driven approach. In: Massacci, F., Wallach, D., Zannone, N. (eds.) ESSoS 2010. LNCS, vol. 5965, pp. 166–181. Springer, Heidelberg (2010)
4. Blanchet, B.: An Efficient Cryptographic Protocol Verifier Based on Prolog Rules. In: 14th IEEE workshop on Computer Security Foundations, p. 82 (2001)
5. Avalle, M., Pironti, A., Sisto, R., Pozza, D.: The Java SPI framework for security protocol implementation. In: Sixth International Conference on Availability, Reliability and Security (ARES), pp. 746–751 (2011)
6. Avalle, M., Pironti, A., Sisto, R.: Formal verification of security protocol implementations: a survey. In: Formal Aspects of Computing (to appear)
7. Bella, G., Massacci, F., Paulson, L.C.: Verifying the SET purchase protocols. J. Autom. Reason. 36(1-2), 5–37 (2006)
8. Nipkow, T., Paulson, L.C., Wenzel, M.T.: Isabelle/HOL. LNCS, vol. 2283. Springer, Heidelberg (2002)
9. Borek, M., Moebius, N., Stenzel, K., Reif, W.: Model-driven development of secure service applications. In: Proceedings of the 35th Annual IEEE Software Engineering Workshop (SEW), pp. 62–71. IEEE (2012)
10. Borek, M., Moebius, N., Stenzel, K., Reif, W.: Model checking of security-critical applications in a model-driven approach. In: Hierons, R.M., Merayo, M.G., Bravetti, M. (eds.) SEFM 2013. LNCS, vol. 8137, pp. 76–90. Springer, Heidelberg (2013)
11. Armando, A., et al.: The AVANTSSAR platform for the automated validation of trust and security of service-oriented architectures. In: Flanagan, C., König, B. (eds.) TACAS 2012. LNCS, vol. 7214, pp. 267–282. Springer, Heidelberg (2012)
12. Jürjens, J.: Developing high-assurance secure systems with UML: a smartcard-based purchase protocol. In: 8th IEEE International Conference on High Assurance Systems Engineering, pp. 231–240 (2004)
13. Gunawan, L.A., Kraemer, F.A., Herrmann, P.: A tool-supported method for the design and implementation of secure distributed applications. In: Erlingsson, Ú., Wieringa, R., Zannone, N. (eds.) ESSoS 2011. LNCS, vol. 6542, pp. 142–155. Springer, Heidelberg (2011)
14. Gunawan, L.A., Herrmann, P.: Compositional verification of application-level security properties. In: Jürjens, J., Livshits, B., Scandariato, R. (eds.) ESSoS 2013. LNCS, vol. 7781, pp. 75–90. Springer, Heidelberg (2013)
15. Dolev, D., Yao, A.C.C.: On the security of public key protocols. IEEE Transactions on Information Theory 29(2), 198–207 (1983)
16. Pozza, D., Sisto, R., Durante, L.: Spi2Java: automatic cryptographic protocol Java code generation from spi calculus. In: 18th International Conference on Advanced Information Networking and Applications, 2004, vol. 1, pp. 400–405 (2004)
17. Pironti, A., Sisto, R.: Provably correct Java implementations of Spi Calculus security protocols specifications. Computers & Security 29, 302–314 (2010)

Fault-Tolerant Non-interference

Filippo Del Tedesco, Alejandro Russo, and David Sands

Chalmers University of Technology, Sweden

Abstract. This paper is about ensuring security in unreliable systems. We study systems which are subject to transient faults – soft errors that cause stored values to be corrupted. The classic problem of fault tolerance is to modify a system so that it works despite a limited number of faults. We introduce a novel variant of this problem. Instead of demanding that the system works despite faults, we simply require that it remains secure: wrong answers may be given but secrets will not be revealed. We develop a software-based technique to achieve this fault-tolerant non-interference property. The method is defined on a simple assembly language, and guarantees security for any assembly program provided as input. The security property is defined on top of a formal model that encompasses both the fault-prone machine and the faulty environment. A precise characterization of the class of programs for which the method guarantees transparency is provided.

1 Introduction and Overview

Transient faults occur in hardware for example when a high-energy particle strikes a transistor, resulting in a spontaneous bit-flip. Such events have been acknowledged as the source of major crashes in server systems [6]. The trend towards lower threshold voltages and tighter noise margins means that susceptibility to transient faults is increasing.

From a security perspective, transient faults (henceforth we will say simply faults) are a known attack vector. For instance, in [7,3,20] a single bit flip, regardless of how is triggered, can compromise the value of a secret key in both public key and authentication systems. In [17] it is shown how a fault (induced by holding a light-bulb near the processor!) triggers a single bit flip in a malicious but well-typed Java applet, causing it (with high probability) to do something which is otherwise impossible for well-typed bytecode: to take over the virtual machine.

Much previous work on *fault tolerance* has studied the preservation of functional behavior or mitigation of faults. For the most part techniques employ wholesale hardware replication, or at least some special-purpose hardware. For the predominantly-software-based techniques, with the exception of [24], most works do not give precise, formal guarantees.

In this work, rather than attempting to preserve full functional behavior in the presence of faults, we consider the novel problem of guaranteeing security: faults may cause a program to go wrong, but even if it goes wrong it should not leak sensitive data, no matter if the code is crafted with malicious intent (cf. [17]). The particular security characterization we study is *non-interference*, a well-established end-to-end information-flow security property which says that public outputs of a program (the *low* security channel) do not reveal anything about its secrets (the *high* security inputs).

J. Jürjens, F. Piessens, and N. Bielova (Eds.): ESSoS 2014, LNCS 8364, pp. 60–76, 2014.

Our approach has two distinguishing features. Firstly, it does not rely on special purpose hardware features (in contrast to [24]), and secondly, it makes its assumptions precise and provides formal guarantees. This latter point distinguishes our approach from software-based techniques used in the large majority of works in fault tolerance which are usually evaluated empirically, often using simulated errors. It should be noted, of course, that our goal is simply to preserve non-interference, and not to detect errors or recover from them.

In the remainder of this section we give an overview of the approach taken in this work to achieve what we called *fault-tolerant non-interference*, and summarize the main results.

The Target System and the Faulty Environment. Transient faults are a feature of hardware, so it makes sense to have an explicit hardware representation. In this paper we consider a single core machine that executes a small set of RISC-like instructions. The machine has registers and two separate memories for code and for data (§ 2.1). We assume the code memory is read-only (ROM), therefore fault-free. This is a standard assumption since memory with error correcting codes is both efficient and commonplace. On the other hand we assume that both registers and data memory are *not* fault-free. This means, in particular, that even the program-counter and hence the control flow can be affected by faults, an assumption in line with most CPU implementations. This is the feature of the system (and systems in general) which makes the problem particularly challenging.

Since we aim for precise guarantees, we assume there is no operating system between programs and the underlying hardware. This choice simplifies the implementation of our method and the security argument. In fact, since the execution of the operating system would be subject to faults, none of its abstractions could be used in a reliable way, and the code would introduce further vulnerabilities.

We assume that the fault environment can simultaneously induce multiple bit-flips in any register or any part of the data memory.

Enforcing Non-interference in the Presence of Transient Faults. Our method enforces security via program transformation. Security is defined in terms of two secrecy levels, *low* for public and *high* for confidential data; low input data may influence the high outputs, but high inputs should not affect the low outputs of the system.

Our transformation combines *Secure Multi-Execution* (SME) [15] [1] with a technique known from Software-based Fault Isolation (SFI) [31] to guarantee that the security property enforced by SME is not compromised by faults.

Consider the system consisting of high and low inputs and outputs represented in Figure 1. The SME version of this system is given in Figure 2. SME deploys two isolated copies of the system, one with responsibility for computing the low outputs, and one with the responsibility of computing the high ones. In our instantiation of this idea, the "system" will be the program to be secured.

A natural approach to implementing SME is to use fair concurrency to compute independently each copy of the system. In our case, the approach has necessarily to

---

[1] Related ideas have appeared elsewhere [27,9,12,5]

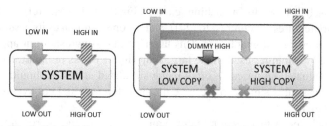

Fig. 1. Original System   **Fig. 2.** Secure Multi-Execution

be more straightforward, since software and hardware supports for concurrency are missing. For this reason, SME is implemented by executing the high copy sequentially after the low one. This mandatory choice makes SME vulnerable to leakage in the presence of faults (§ 2.2-2.3). In particular:

- during execution of the low copy, a fault in the value of a pointer stored in a register could cause the high data to be loaded instead of low;
- during the execution of the high copy, a fault in the program counter can cause the control-flow to transfer to the low copy, but in a state where the registers might contain arbitrary high data.

In both of these scenarios, the low copy of the code gains access to the high data. The attacker's ability to take advantage of this may depend on the structure of the code, or the attacker's ability to recognize a leaked secret independently of the code. Nevertheless, to construct a general security mechanism based on SME, we must protect against the situations enumerated above.

A typical assumption in the analysis of fault tolerance mechanisms is the occurrence of a single fault. Similarly, we strengthen SME so that it can cope with at most some small fixed number of faults (§ 3.3). The key to preserving the strong isolation provided by SME, in the presence of up to F faults, is to

- (§3.1) separate the address space of the high and low variants of the code, and the data memory addresses over which they operate so that the addresses of the respective parts have a hamming distance[2] greater than F
- (§3.2) add address masking code, in the style of SFI, around load and jump instructions to mask the address value so that it is forced within in a safe range.

As for the original SME, our method guarantees isolation between *low* and *high* components in a language-independent manner, since systems are treated as black boxes; moreover, such isolation remains unaltered even if F faults occur during the execution. Our method guarantees *transparency* as well: if the original system had no information leaks between high inputs and low outputs, and no faults occur in the execution, then the modified system will produce the same values on the low and high channels as the original system (since the dummy high input will have no influence on the computation).

Results. For security, we formalize the semantics of the machine (§ 4.1) and precisely specify our assumptions about which faults can occur (§ 4.2). From this we formulate

---

[2] The number of positions for which corresponding bits of two equally sized binary words differ.

a suitable notion of non-interference (§ 4.3), where we tackle the problem that faults, when modeled as nondeterminism, can mask information flows.

Surprisingly, security is established with no semantic assumptions about the code itself. In order to guarantee transparency we need "reasonable" semantic invariants (§ 5) on memory utilization and control flow modifications performed by the source program.

2 Transient Fault Based Attacks on SME

This section illustrates the syntax of assembly programs and the inadequacy of a naive SME implementation in the presence of faults.

2.1 Syntax

Data manipulated by assembly programs are in the set Val, which is defined as the disjoint union of $\mathbb{W} \cup Ptr \cup Lab \cup DReg$. The set $\mathbb{W}$ corresponds to numeric constants, defined as machine words of n bits. Pointers to data memory, from the set $Ptr \overset{\text{def}}{=} \{ptr\, v \mid v \in \mathbb{W}\}$, are defined as tagged machine words to keep them separated from elements in $\mathbb{W}$. We assume an infinite set of labels Lab, representing targets of jump instructions, and a finite set of general purpose registers $DReg$.

$$
\begin{aligned}
I &::= [l:]B \text{ such that } l \in Lab \\
B &::= \text{load } r\, v \mid \text{store } v\, r \mid \text{jmp } v \quad \mid \text{jnz } v\, r \mid \\
&\quad \text{nop} \quad \mid \text{move } r\, v \mid BinOp\, r\, v \mid \text{out } ch\, r \\
BinOp &::= \text{add} \quad \mid \text{or} \\
P &::= \epsilon \mid I :: P
\end{aligned}
$$

Fig. 3. Assembly programs syntax

Figure 3 shows the syntax for assembly programs. We consider that every instruction I could be optionally labeled. Instruction load $r\, v$ accesses the data memory and writes the value pointed by v into register r. The corresponding store $v\, r$ instruction writes the content of r into the data memory address v. Instruction jmp v causes the control-flow to transfer to the instruction labeled as v. Instruction jnz $v\, r$ performs the jump only if the content of register r is nonzero. Instruction move $r\, v$ copies the value v into register r. $BinOp$ stands for a family of binary operators that combine values in r and v and store the result in r. A minimal such family contains an or instruction and an add instruction. The or instruction performs the logic or operation between constants in r and v; the add instruction adds the unsigned constant v to the value contained in register r, which can either be a constant or a memory pointer. All instructions presented so far are either indirect, when v is in $DReg$, or direct when v is in $Val \setminus DReg$. Instruction nop performs no computation. Instruction out $ch\, r$ outputs the constant contained in r into the channel ch. Output channels are in the set $Out = \{low, high\}$.

Programs are defined as lists of instructions P. We denote the set of labels contained in a program as $lab(P)$. We require programs to be well-formed, namely not having two instruction bodies labeled in the same way. Given two programs P and P', we define program composition $P ++ P'$ as list concatenation, provided that $lab(P) \cap lab(P') = \{\,\}$.

2.2 Direct Control Flow and Memory Faults

We describe how faults can induce secret leakages in SME-programs. Consider Figure 4, in which an assembly program and the memory M on which it is executed are presented. Observe that M contains both a public value pub and a secret sec. The program P is intuitively secure. The first move instruction writes the memory pointer pub_p to register r_1. Then the public value pub is loaded in r_2, and sec_p overwrites pub_p in r_1. Finally, pub is output on the low channel via the last out instruction.

Since program P is secure, its SME version, written $sme(P)$, is also secure [15]. Figure 5 shows the code of $sme(P)$ and the corresponding memory. The transformed program consists of the two copies of program P, named P_{low} and P_{high}, responsible for computing public and secret values, respectively. The memory is divided into the segments μ_{low} and μ_{high} in such a way that the code in P_{low} only refers to μ_{low} and the code

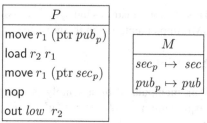

in P_{high} only to μ_{high}. The segment μ_{low} contains the dummy value zero ($sec'_p \mapsto 0$) instead of the secret value sec, while instructions for public outputs are replaced by nop in P_{high}. Clearly, $sme(P)$ preserves confidentiality.

We proceed to describe how a single bit flip is enough to jeopardize the security guarantees of $sme(P)$. In a machine execution, it could be possible for sec_p and pub'_p to be located at the memory addresses 000 and 100, respectively. It is then possible for pub'_p to be converted to sec_p by a single bit flip. As a consequence, the secret value sec could be loaded into r_2 by the second instruction in P_{low}, which in turn would send it on a low channel.

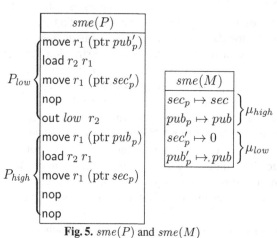

Fig. 5. $sme(P)$ and $sme(M)$

Bit flips in the program counter are problematic as well. Suppose the execution goes through P_{low} and completes the first nop in P_{high} without faults. At this point, the program counter contains the value 9 (1001 in binary), i.e., it points to the last instruction of P_{high}, and the register r_1 contains the pointer sec_p. However, just before the last instruction of P_{high} is executed, a bit flip in the first bit of the program counter can move the execution back to 0001, i.e., the second instruction

of P_{low}. Since this occurs while r_1 contains sec_p, it is possible for P_{low} to have access to sec, and leak it on the low channel.

The scenarios described above suggest that in order to guarantee security in a faulty context, SME has to separate P_{low}, P_{high}, μ_{low}, and μ_{high} in a way that tolerates bit flips in memory pointers or in the program counter, as discussed in Section 3.1.

2.3 Indirect Control Flow and Memory Faults

Faults can induce arbitrary computations *within* P_{low} and P_{high}. Although we do not attempt to preserve functional correctness in the presence of faults, performing arbitrary computations in a SME scenario has important security implications.

Consider the fragment of *low* code in Figure 6. Alterations in the program counter could bypass the initialization of r_1 to ptr pub_p and use an arbitrary value • as memory pointer. Hence, regardless how μ_{low} and μ_{high} are spread out in memory, it would be still possible for a pointer in P_{low} to refer to values in μ_{high}. This situation can clearly jeopardize the security guarantees of SME. Observe that arbitrary computations on P_{high}'s memory pointers do not present any security risks. After all, it is

```
move r₁ •
move r₁ (ptr pubₚ)
nop
load r₂ r₁
```

Fig. 6. *low* code

secure for P_{high} to access μ_{low}. However, perturbations in P_{high}'s control flow impose other danger.

When P_{high} is executed, faults in the program counter could induce arbitrary values to be used as jump targets. When this is the case, the control flow can be moved from P_{high} back to P_{low}, regardless how P_{low} and P_{high} are located in memory. Since secret data is often loaded into registers by P_{high}, this type of jumps presents a security risk. Observe that there is no risk for arbitrary computations to trigger jumps from P_{low} to P_{high}.

In Section 3.2 we propose to use instrumentations for instructions load, jmp, and jnz so that leaks can be prevented even in the presence of arbitrary computations.

3 Fault-Tolerant Secure Multi-execution

We present a version of SME capable of preserving confidentiality of high inputs even in a faulty environment. Our technique relies on spreading out code (P_{low} and P_{high}) and memory (μ_{low} and μ_{high}) as well as instrumenting instructions related to memory access and jumps.

3.1 Fault-Tolerant Layout for Code and Memory

Fault tolerance always involves some kind of redundancy. In our case we will use the first $F + 1$ bits of every n-bit address exclusively for keeping the hamming distance between P_{low} and P_{high}, and between μ_{low} and μ_{high}, to at least $F + 1$.

Let $distance(u, v)$ be the hamming distance between two words u and v. We will say that two words are F-separate whenever their hamming distance is greater than F.

We will work with programs for which both their size, and their run-time memory footprint, is roughly in the range $[0, 2^{n-(F+1)} - 1]$ (the exact range may be slightly smaller than this and can be calculated after some additional instructions have been

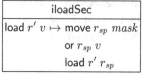

iloadSec
load r' $v \mapsto$ move r_{sp} $mask$
or r_{sp} v
load r' r_{sp}

Fig. 7. Securing load

ijmpSec
jmp $v \mapsto$ move r_{sp} $mask$
or r_{sp} v
jmp r_{sp}

Fig. 8. Securing jmp

ijnzSec
jnz v $r' \mapsto$ move r_{sp} $mask$
or r_{sp} v
jnz r_{sp} r'

Fig. 9. Securing jnz

inserted into the code according to the transformation described in the next subsection). The remaining bits of the address spaces (code and data memory) are reserved for our fault tolerance mechanism.

Let $mask$ denote the word with $F + 1$ leading 1s followed by $n - (F + 1)$ zeros.

The idea is that any address in the range $[b, t]$ (where $b < t < 2^{n-(F+1)}$) is F-separate from any address in the range $[b + mask, t + mask]$.

If μ_{high} occupies the memory addresses in the interval $[0, t]$ then we ensure that μ_{low} uses the range $[mask, t + mask]$. This clearly gives F-separation between μ_{low} and μ_{high} and thus avoids leaks due to faults in pointers handled by P_{low} (see Section 2.2).

For achieving a similar separation between P_{high} from P_{low} we add some code padding between the two copies of P such that the first instruction of P_{high} is at the ROM address $mask$. This guarantees F-separation between the addresses of instructions in P_{low} and P_{high} and thereby avoids leak due to direct faults in the program counter while executing P_{high} (see Section 2.2).

3.2 Control Flow Integrity

Faults can break the control-flow integrity of the program, causing it, for example, to jump to an arbitrary address. The two problematic instances of this problem are when (i) P_{low} loads from an address in μ_{high}, and (ii) when the destination of a jump in P_{high} points to P_{low}. We mitigate these cases using a technique which turns out to be very similar to the sandboxing approach in software-based fault isolation [31]: we mask the addresses so that they are always within a safe range. This is achieved in case (i) by transforming load instructions, and in case (ii) by transforming jmp and jnz instructions, as shown in Figures 7 to 9.

Note that for this to work we need one spare general purpose register r_{sp} – i.e., one which is not used by the original program P.

3.3 Formal Definition of Fault-Tolerant SME

Figure 10 summarizes the process of generating our fault-tolerant version of SME as a program transformation. SME reworks an assembly program P into two secure variants P_{low} and P_{high}. This requires modifications to the internal behavior of program P. The transformation consists of several steps. To obtain P_{high} from P, we first replace the instructions to write data into public channels by nops. This is done by the function o_{low}, which generates an intermediate result P'_{high}. Function jnzSec ∘ jmpSec (the symbol ∘ denotes function composition) instruments jmp and jnz instructions by applying functions in Figures 8 and 9 to the entire program.

Obtaining P_{low} is a bit more involved. It requires offsetting every pointer appearing in P by $mask$ so that P_{low} refers to μ_{low} (function offset$_{mask}$). Additionally, the transformation renames instruction labels to avoid name clashes with P_{high} (function lab$_P$), as well as suppressing instructions performing outputs in high channels (function o$_{high}$).

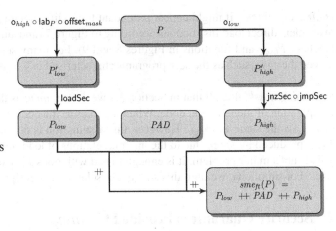

Fig. 10. Fault-tolerant SME code transformation (sme_{ft})

The instrumentation of load is done by function loadSec (based on the auxiliary function in Figure 7), thus finally obtaining P_{low}. Once P_{low} and P_{high} are obtained, in order for F-separation to hold between them, the transformation adds some padding code, named PAD. All instructions in PAD are jumps to the first instruction of P_{high}, and the length of PAD guarantees the first instruction of P_{high} is located at the address $mask$ (recall Section 3.1).

Initial memory configuration. Consider the initial memory M for P in Figure 11. We assume that the program uses the memory interval $\mu = [0, t]$, where the first s words in M are secrets (labeled $high_{in}$), the subsequent words are public values (low_{in}) and the rest is uninitialized (in white). We require s to be within the range $[0, 2^{n-(F+1)} - 1]$ to ensure the separation between μ_{high} and μ_{low} is possible (Section 3.1).

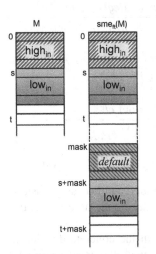

We also require that M only contains values from $\mathbb{W}$. The security of the method does not depend on this assumption, but for the transformation to preserve the non-faulty behavior of secure runs of the program we will need such requirement on input. We return to this issue in Section 5. Under these assumptions, the initial memory for $sme_{ft}(P)$, which we denote by $sme_{ft}(M)$, corresponds to the right side of Figure 11. Notice that μ_{high}, the portion of the memory to be used by P_{high}, is the same as μ, whereas P_{low} will use μ_{low} which is located in the memory interval $[mask, t + mask]$. In μ_{low} the words representing the secret are initialized to a default value (marked "default" in the figure). For the sake of simplicity, we do not require $sme_{ft}(P)$ to take care of memory rearrangement itself – we assume the preparation of $sme_{ft}(M)$ is external to SME. We assume initial registers to be all uninitialized for P, therefore they will be uninitialized for $sme_{ft}(P)$ as well.

Fig. 11. Initial memory M and transformed version $sme_{ft}(M)$

Optimizing sme_ft. It might appear redundant to modify memory pointers in P_{low} *and* instrument direct load instructions according to Figure 7 (and similarly for control flow labels in P_{high} and functions in Figures 8 and 9). For many sensible programs this is indeed the case, such as the *safe* programs characterised in § 5.

Redefining mask. Recall that in Section 3.1 we define *mask* as the mask used to obtain F-separation of memory and code. When it comes to the code, we assume that the size of P_{low} is the same as P_{high}. However, this assumption is no longer true for P_{low} and P_{high} produced by *sme_ft* due to the instrumentations of load, jmp and jnz instructions. This is not a major problem. It is enough to pad with nops P_{low} or P_{high} to match their sizes. For simplicity, we omit this step in our schematic description.

4 Security Guarantees Provided by *sme_ft*

In this section we state the security property bestowed by *sme_ft* on transformed programs. To do this we define a formal semantics for the RISC machine; extend it to model faults; define non-interference for faulty runs; state the security theorem: any program transformed by *sme_ft* corresponds to a machine program which is non-interfering for runs with no more than F faults. For space reasons most of the details are not given here; we refer to the full version [13].

4.1 Semantics

To give a precise semantics to faults we need to work at the level of concrete programs, i.e., *machine code*, which are lists of concrete instructions. Compared to assembly instructions from Figure 3, concrete instructions are not labeled, and their arguments are register names or machine words. This formalization of machine code is sufficiently concrete to describe the class of faults we wish to model. In particular, a concrete encoding of the register names is not made

$$\text{DLoad} \; \frac{P(pc) = \text{load}_d \; r \; w}{\langle P, Reg, M \rangle \xrightarrow{\tau} \langle P, Reg^+[r \mapsto M(w)], M \rangle}$$

$$\text{DAdd} \; \frac{P(pc) = \text{add}_d \; r \; w \quad Reg(r) + w = w'}{\langle P, Reg, M \rangle \xrightarrow{\tau} \langle P, Reg^+[r \mapsto w'], M \rangle}$$

$$\text{DJnz-A} \; \frac{P(pc) = \text{jnz}_d \; w \; r \quad Reg(r) \neq 0}{\langle P, Reg, M \rangle \xrightarrow{\tau} \langle P, Reg[pc \mapsto w], M \rangle}$$

$$\text{Out} \; \frac{P(pc) = \text{out} \; ch \; r}{\langle P, Reg, M \rangle \xrightarrow{ch!Reg(r)} \langle P, Reg^+, M \rangle}$$

Fig. 12. Concrete Semantics (selected rules)

explicit because we do not consider faults in the code memory, and because registers are not addressable indirectly. We sometimes write $P(i)$ to denote the ith concrete instruction in the instruction list P.

Most assembly instructions have two explicit versions in the concrete domain: a *direct* version, such as $\text{load}_d \; r \; w$ which loads the value contained at memory address w into the register r, and an *indirect* version, such as $\text{load}_i \; r \; r'$ which fetches the memory address of the data to be loaded from register r'. There are two exceptions to this: the nop instruction, which does not require any parameter, and the out instruction, which has no direct formulation. Observe that, similarly to register names, channel names are not encoded.

Assembly programs are converted to concrete ones by the function loader. The function converts abstract values *Val* into machine words. In particular this amounts to stripping the pointer tag away from the pointers, and resolving code labels to ROM addresses. The function loader is also responsible for mapping all abstract instructions into their direct or indirect versions. The details are straightforward and not presented here [13].

Configurations of the concrete machine are given by a triple $\langle P, Reg, M \rangle$, where P is the concrete program, $Reg \in DReg \cup \{pc\} \to \mathbb{W}$ is the *(Concrete) Register Bank* and $M \in \mathbb{W} \to \mathbb{W}$ is the *(Concrete) Data Memory*.

The fault-free semantics of concrete programs is given as a labeled transition system. The labels on transitions indicate the observable output of each clocked machine step, and are either τ, a label marking just the passage of time, or an output label, indicating a word output on a specific channel. All labels are in $Act = \{low!w | w \in \mathbb{W}\} \cup \{high!w | w \in \mathbb{W}\} \cup \{\tau\}$. A representative selection of reduction rules for the concrete machine are presented in Figure 12. We use Reg^+ as a shorthand for $Reg[pc \mapsto Reg(pc) + 1]$ and we abbreviate $P(Reg(pc))$ as $P(pc)$. Modelling instructions as consecutive words implies that it is impossible to jump to an address which is not aligned with the beginning of an instruction; this assumption corresponds to the implementation of simpler RISC architectures such as ARM versions 1 and 2.

4.2 Modeling Faults

Our aim will be to describe the overall behavior of a fault-prone system as simply as we can, while still permitting reasoning about non-interference. The core idea is to model the transitions of the system in the presence of faults with a labeled transition system obtained by interleaving the machine transitions with a nondeterministic flipping of zero or more bits. As described previously, the fault-prone bits of the machine are any of the register bits, and any bits in the data memory.

We need some notation to talk about bit flips. Recall machine words are n bits long. Let us define the set of *locations* at which a fault may occur as:

$$Loc \stackrel{\text{def}}{=} \{(r, i) \mid r \in DReg \cup \{pc\}, i \in \{1, \dots, n\}\} \cup \{(k, i) \mid k \in \mathbb{W}, i \in \{1, \dots, n\}\}$$

For a machine configuration C and location $l \in Loc$ we will write $C[l]$ to denote the value of the bit specified by l in C; for any $b \in \{0, 1\}$ we write $C[l \mapsto b]$ to denote the configuration obtained from C by updating the location l to b.

Let L range over the (possibly empty) subsets of locations. We express bit flips in the values of a given subset L of locations by using the function flip defined as $\text{flip}(C, L) = C[l \mapsto \neg C[l], l \in L]$, which flips every bit of locations L in the machine configuration C.

We can now define faulty systems with labeled transitions ($\stackrel{a}{\rightsquigarrow}$, $a \in Act$) with the transition rule to the right. It can be seen from the rule that our fault model assumes

$$\frac{\text{flip}(C, L) \stackrel{a}{\rightarrow} C' \quad L \subseteq Loc}{C \stackrel{a}{\rightsquigarrow} C'}$$

that the transitions of the system are instantaneous (a common assumption, but a potential source of inaccuracy – a point we return to in the conclusions). The fact that faults can occur between transitions is modeled by allowing any fault to occur before any transition of the system is taken. The number of faults occurring in a given transition is $|L|$, and is not constrained in this rule, but will be constrained at the level of *runs*.

4.3   Fault-Tolerant Non-interference

This section formalizes the confidentiality guarantees of our approach in the presence of faults.

Since the faulty system is nondeterministic, one might consider a simple *possibilistic* notion of non-interference — secret values should not influence the *set of possible public outputs* of the faulty system. This notion is not adequate because unfortunately errors might occur anywhere, in particular on public values, therefore any program is capable to produce any possible output!

This is an instance of a known weakness of possibilistic non-interference [18,22]. A standard fix is to adopt a *probabilistic* notion of non-interference – the probability distribution of public outputs is unaffected by the secrets in the presence of errors – assuming an attacker can perform probability measures. In this paper, however, we adopt a different approach: we permit the attacker to observe *exactly when and where faults occur* in a given run, along with output events in the low channel and the passage of time. This model leads to a security definition which seems stronger than the probabilistic one, but in fact we have shown [14] that the two notions are equivalent for the computational model considered here.

We start concretising the attacker's view of a system by defining function $low \in Act \to \{low!w | w \in \mathbb{W}\} \cup \{\tau\}$. More precisely, $low(a)$ returns a if $a = low!w$, and returns τ otherwise. Now we can define the semantics of the faulty system from the attacker's perspective as a labeled transition system given by the following transition rules:

$$\text{Step } \frac{\text{flip}(C, L) \xrightarrow{a} C'}{C \xrightarrow{L, low(a)} C'} \qquad \text{Stuck-1 } \frac{\text{flip}(C, L) \nrightarrow}{C \xrightarrow{L, \tau} \text{flip}(C, L)} \qquad \text{Stuck-2 } \frac{C \nrightarrow}{C \xrightarrow{L, \tau} C}$$

The attacker observations imply that termination of the system is not directly observable and that once a system reaches a stuck configuration, faults have no further effect.

We can now state our security condition. We say a machine configuration is *initial* if (i) $Reg(pc) = 0$, (ii) $Reg(r_{sp}) = 2^n - 1$ (so it never points to low code/high data), and (iii) secrets are stored in the first s words of the memory (Figure 11).

We say two initial configurations C and C' are *low equivalent*, written as $C =_{low} C'$ if they differ, at most, on the first s words of the heap.

We say that a sequence $\sigma = L_0, a_0, \ldots L_{n-1}, a_{n-1}$ is a *low run* of a system state C_0 whenever there exist states $C_1, \ldots, C_n$ such that $C_i \xrightarrow{L_i, a_i} C_{i+1}$ for all $i \in \{0, \ldots, n-1\}$. The number of faults exhibited by σ is $\Sigma_{i=0}^{n-1}|L_i|$.

Definition 1 (*F*-**Fault-Tolerant Non-interference**). *An initial configuration C is F-fault-tolerant non-interfering if for all initial configurations C' such that $C =_{low} C'$, the set of low runs exhibiting no more than F faults are the same for C and C'.*

We say that an assembly program P is F-fault-tolerant non-interfering if all initial configurations relative to P, namely $\langle \mathsf{loader}(P), Reg, M \rangle$ are F-fault-tolerant non-interfering.

Theorem 1 (**Non-interference induced by** sme_{ft}). *If $sme_{ft}(P) = P'$ then P' is F-Fault-tolerant non-interfering.*

The theorem is proved by showing that (i) all memory accesses in P_{low} are performed towards addresses that are F-separate from μ_{high} and (ii) once the computation reaches P_{high} it cannot be moved back to P_{low}.

Both properties depends on the layout of code and data memory, together with on the invariant property on r_{sp}. In particular we can show that in the absence of faults, the value contained in r_{sp} is in the range $[mask, 2^n - 1]$, whereas in the presence of faults the content of r_{sp} is never in the range $[0, 2^{n-(F+1)} - 1]$. For a detailed proof refer to [13].

Definition 1 is both termination and (logical) timing sensitive: we require that any two runs of the system (that exhibit at most F faults) correspond to the same sequence of observable events, regardless of secret data. Not only output values must be the same, but the instant in which they occur must coincide as well. Hence, Theorem 1 guarantees that our transformation technique can secure all programs whose timing and termination behavior can induce leaks.

5 Transparency Guarantees Provided by sme_{ft}

We have shown that the transformed programs meet the goal of non-interference in the presence of faults. We have done so with no semantic assumptions about the code itself. The only *syntactic* assumptions are on the size of the code, which is required to be small enough to accommodate the transformation in the ROM, on the amount of secret data in the initial memory, and on the registers utilization – we require at least one spare register.

Does the transformation sme_{ft} preserve the behavior of programs? The answer, in general, is no. Firstly, programs which are intrinsically insecure exhibit a different behavior under standard SME. This alteration in the semantics is done in order to enforce confidentiality. It could be said that "software faults", i.e., instructions leaking secret data, are being mitigated by SME. However, even when the original program is secure, our transformation modifies the size and layout of the original program and the absolute location of data in memory. In general machine code programs can be sensitive to such transformation, and behave in an arbitrarily different way.

For this reason, transparency guarantees can be given only for programs which are "sensible" and secure for fault-free runs. We consider a program "sensible" when it is *safe* and *bounded*. A program is *safe* when, roughly speaking, it is not sensitive to the absolute addresses of its instructions in the ROM, or the absolute addresses of the memory that it accesses. A program is *bounded* when there is a known upper bound on the region of memory that it will address.

For any "sensible" program, the following theorem holds:

Theorem 2 (Transparency). *(informal statement) Let P be a non-interfering, "sensible" assembly program. If the low copy P_{low} always terminates, then the SME transformed program $sme_{ft}(P)$ yields the same sequence of values on each of the respective output channels as P for any fault-free run.*

A formal account of Theorem 2 (and its proof) can be found in the full version of the paper [13].

In this work the characterization of safe and bounded programs is obtained via an abstract machine for the language. The abstract machine characterises those programs which never exhibit certain "bad" behaviours. This is in the same spirit as e.g. Leroy's compiler correctness proof [21]. We expect that any program correctly compiled from a strongly-typed high level language, and which has a statically known memory footprint, will be a safe and bounded program. To give these guarantees formally one could use a verified compiler, or it could be achieved by compiling to a typed version of our assembly language (see, for example, [23]) which ensures that the produced code is safe and bounded. However, these endeavours lie outside the scope of the present paper.

Notice that for Theorem 2 to hold we require the low copy of the source program to terminate on all input. This means that, in general, transparency does not hold for programs that are nonterminating by construction (e.g. server applications). However, this does not compromise security: Theorem 1 holds for this class of programs as well.

6   Related Work

Language Based Dependability. The use of application-layer techniques for achieving fault tolerance have been widely studied. De Florio and Blondia survey the field [16] and classify the various ways in which fault tolerance can be added, and what kind of faults are supported. Notably, none of the techniques surveyed at that time either deal with tolerance with respect to security properties, or with techniques that give precise semantic guarantees.

More recently, Project Zap [1] has applied language based techniques to transient faults modeling and analysis with the goal of providing formally verifiable dependability methods. The closest to our work in the Zap series is the work on fault-tolerant typed assembly language of Perry et al [24]. We use an abstract machine to characterize the class of programs for which our method is applicable. Our characterization is more liberal than a typical typed assembly language, but a typed assembly language could nevertheless be used as a sound method to prove that a program is safe and bounded. Both in that work and in ours, transient faults have a semantic interpretation as nondeterministic transitions that can happen at anytime and anywhere in the faulty hardware. Since we do not aim at functional correctness preservation, we can be more liberal in the class of faults we admit (more than one bit flipped at a time) and in the hardware components the concrete machine operates on. In [25] the attention is solely focused on detecting control flow modifications induced by transient faults. The method, unlike

[24], is purely software based. However, detectability is possible only for programs that obey a strict control-flow discipline, and under the assumption that at most a single bit flip occurs. Once again, our ability to cope with a bigger class of control flow errors comes from the fact that we aim for a weaker property; arbitrary control flow alterations inside P_{low} or P_{high} executions do not pose security threats.

Fault Isolation Techniques. As mentioned previously, the techniques we use to mask addresses to prevent dangerous loads and jumps can be found in the software-based techniques for fault isolation (SFI) introduced by Wahbe *et al* [31] for sandboxing untrusted code. A similar address-masking technique is used in [10] for mitigating the effects of transient faults. Also, principles from SFI are also implemented in [2], where the authors define a method to prevent an active attacker from corrupting the control flow integrity of a program.

It should be noted, however, that the "faults" targeted by SFI are those caused by buggy/malicious code or data. The SFI techniques, in isolation, are able to protect from the effects of some but not all of the transient faults studied here.

What we said for software based methods also hold for sandboxing techniques using special operating system or hardware features – they are not designed for and do not protect against all transient faults, and may increase the attack surface (via increased code or by relying on special purpose registers).

Fault Tolerance vs Non-Interference. As we have shown in our result, fault tolerance and non-interference present interesting connections, and we believe that our combination is a novel one. However other connections between the two concepts have been noted in a number of other works.

The *Strong Security* notion introduced by Sabelfeld and Sands in [29] for multi-threaded programs is shown to be strong enough to guarantee an unrestricted form of fault-tolerant non-interference in [14], providing a more restrictive class of transient faults are considered (faults cannot corrupt the control flow integrity). In a similar way, programs that are secure according to the definition in [28], an extension of [29] to distributed systems, can be shown to retain security regardless of faults occurring in network communications. It is not surprising that both cases cannot cope against faults in the control flow since, as we have shown in Section 2, control flow alterations introduce completely unexpected information flows.

Another interesting aspects of the comparison between fault tolerance and non-interference was observed by Weber [33]. In this work the author explores a non-interference-like characterisation of fault tolerance in terms of program semantics. A more general view on the connection between enforcement mechanisms for information flow properties and dependability goals is proposed by Rushby [26]. Overall the techniques used in the present work can be understood in terms of the general partitioning mechanisms described by Rushby. In particular what Rushby calls *spatial partitioning* corresponds to our separation of memory addresses (albeit within the same physical memory); *temporal partitioning* characterises what we achieve by ensuring that low events happen before high events, since this ensures that the timing of high events cannot influence low events.

Security Preservation in the Presence of Transient Faults. Our method guarantees that security of programs, expressed in terms of F-Fault-Tolerant Non-interference, is preserved even when a limited number of bit flips occur. Other forms of security preservation in faulty environments have been studied, particularly in cryptography.

In [4] authors illustrate several transient-fault based attacks on RSA and Discrete Logarithms cryptographic schemes, together with software countermeasures. Such protection mechanisms involve either some form of replication (they basically require to repeat the computation twice and check the result for fault detection) or a more intensive usage of randomness in the intermediate stages of cryptographic operations to increase the unpredictability of the result.

In [11] authors show how the parameters of an elliptic curve cryptosystem can be compromised by transient faults, and illustrate how a comparison mechanism is sufficient to prevent the attack from being successful. In particular the method compares the working copies of said parameters (located in a faulty hardware component) to their original counterparts (stored in fault-free hardware) in several stages of the computation. Canetti et al [8] discuss security in the presence of transient faults for cryptographic protocol implementations where they focus on how random number generation is used in the code. Harrison et al consider [19] a "confinement problem in the presence of faults", but their work concerns faults in the sense of abnormal termination of software, and the proper confinement thereof.

7   Conclusion and Further Work

We have presented a technique to make programs secure despite a small number of faults, and characterized when the method preserves the behavior of programs. The problem we study is itself novel, and relative to the faults we model, it is notable that our technique does not demand special hardware, and is capable of tolerating multi-bit errors.

Perhaps the main weakness of the present work is the fault model itself. While we model faults in all the main state elements of the machine, we do not model faults in lower-level structures, such as pipelines or in the combinatorial circuits. This shortcoming seems to be shared with much work on fault tolerance (although we do, at least, model faults in the program counter) – in particular works which focus on fault injection e.g. [30]. One might speculate that many faults occurring at the lower level of abstraction are adequately modeled by flipping a few bits in a register, but there seems to be little work to verify this. One of them, by Wang *et al* [32], suggests that lower-level faults are notably rare.

A precise account about the efficiency of our approach is left for further work. An approximate estimation of the overhead can be determined by considering that the system is basically run twice, and all the load and jump instructions are expanded in macros of three instructions each.

Acknowledgment. Many thanks to Johan Karlsson, Ioannis Sourdis, Georgi Gaydadjiev, Arshad Jhumka and the anonymous referees for useful comments and observations. This work was partially financed by grants from the Swedish research agencies VR and SSF, and the European Commission EC FP7-ICT-STREP WebSand project.

References

1. The zap project, http://sip.cs.princeton.edu/projects/zap/ (accessed: February 20, 2013)
2. Abadi, M., Budiu, M., Erlingsson, U., Ligatti, J.: Control-flow integrity. In: Proceedings of the 12th ACM Conference on Computer and Communications Security, CCS 2005, pp. 340–353. ACM, New York (2005),
   http://doi.acm.org/10.1145/1102120.1102165
3. Aumüller, C., Bier, P., Fischer, W., Hofreiter, P., Seifert, J.P.: Fault attacks on rsa with crt: Concrete results and practical countermeasures. In: Kaliski Jr., B.S., Koç, Ç.K., Paar, C. (eds.) CHES 2002. LNCS, vol. 2523, pp. 260–275. Springer, Heidelberg (2003)
4. Bao, F., Deng, R., Han, Y., Jeng, A., Narasimhalu, A., Ngair, T.: Breaking public key cryptosystems on tamper resistant devices in the presence of transient faults. In: Christianson, B., Crispo, B., Lomas, M., Roe, M. (eds.) Security Protocols 1997. LNCS, vol. 1361, pp. 115–124. Springer, Heidelberg (1998)
5. Barthe, G., Crespo, J.M., Devriese, D., Piessens, F., Rivas, E.: Secure multi-execution through static program transformation. In: Giese, H., Rosu, G. (eds.) FORTE/FMOODS 2012. LNCS, vol. 7273, pp. 186–202. Springer, Heidelberg (2012)
6. Baumann, R.: Radiation-induced soft errors in advanced semiconductor technologies. IEEE Transactions on Device and Materials Reliability 5(3), 305–316 (2005)
7. Boneh, D., DeMillo, R.A., Lipton, R.J.: On the importance of eliminating errors in cryptographic computations. Journal of Cryptology 14, 101–119 (2001)
8. Canetti, R., Herzberg, A.: Maintaining security in the presence of transient faults. In: Desmedt, Y.G. (ed.) Advances in Cryptology - CRYPTO 1994. LNCS, vol. 839, pp. 425–438. Springer, Heidelberg (1994)
9. Capizzi, R., Longo, A., Venkatakrishnan, V.N., Sistla, A.P.: Preventing information leaks through shadow executions. In: Proceedings of the 2008 Annual Computer Security Applications Conference, ACSAC 2008. IEEE Computer Society (2008)
10. Chang, J., Reis, G., August, D.: Automatic instruction-level software-only recovery. In: DSN 2006, pp. 83–92 (2006)
11. Ciet, M., Joye, M.: Elliptic curve cryptosystems in the presence of permanent and transient faults. Des. Codes Cryptography 36(1), 33–43 (2005)
12. Cristiá, M., Mata, P.: Runtime enforcement of noninterference by duplicating processes and their memories. In: WSEGI 2009, Argentina. 38 JAIIO (2009)
13. Del Tedesco, F., Russo, A., Sands, D.: Fault tolerant non-interference (extended version) (2013), http://www.cse.chalmers.se/~tedesco/papers/essos14.pdf
14. Del Tedesco, F., Russo, A., Sands, D.: A theory of fault tolerance noninterference (preliminary) (2013)
15. Devriese, D., Piessens, F.: Noninterference through secure multi-execution. In: Proc. of the 2010 IEEE Symposium on Security and Privacy, SP 2010. IEEE Computer Society (2010)
16. Florio, V.D., Blondia, C.: A survey of linguistic structures for application-level fault tolerance. ACM Comput. Surv. 40(2) (2008)
17. Govindavajhala, S., Appel, A.W.: Using memory errors to attack a virtual machine. In: SP 2003, IEEE Computer Society, Washington, DC (2003)
18. Gray, J.W., Probabilistic, I.: interference. In: Proceedings of the 1990 IEEE Computer Society Symposium on Research in Security and Privacy, pp. 170–179 (1990)
19. Harrison, W.L., Procter, A., Allwein, G.: The confinement problem in the presence of faults. In: Aoki, T., Taguchi, K. (eds.) ICFEM 2012. LNCS, vol. 7635, pp. 182–197. Springer, Heidelberg (2012)

20. Kim, C., Quisquater, J.J.: Fault attacks for crt based rsa: New attacks, new results, and new countermeasures. In: Sauveron, D., Markantonakis, K., Bilas, A., Quisquater, J.-J. (eds.) WISTP 2007. LNCS, vol. 4462, pp. 215–228. Springer, Heidelberg (2007)
21. Leroy, X.: A formally verified compiler back-end. J. Autom. Reason. 43(4), 363–446 (2009), http://dx.doi.org/10.1007/s10817-009-9155-4
22. McLean, J.: Security models and information flow. In: Proc. IEEE Symposium on Security and Privacy, pp. 180–187. IEEE Computer Society Press (1990)
23. Morrisett, G., Walker, D., Crary, K., Glew, N.: From system f to typed assembly language. ACM Trans. Program. Lang. Syst. 21(3), 527–568 (1999)
24. Perry, F., Mackey, L., Reis, G.A., Ligatti, J., August, D.I., Walker, D.: Fault-tolerant typed assembly language. In: Proceedings of the ACM SIGPLAN Conference on Programming Language Design and Implementation, pp. 42–53. ACM, New York (2007)
25. Perry, F., Fisher, K.: Reasoning about control flow in the presence of transient faults. In: Alpuente, M., Vidal, G. (eds.) SAS 2008. LNCS, vol. 5079, pp. 332–346. Springer, Heidelberg (2008)
26. Rushby, J.: Partitioning for safety and security: Requirements, mechanisms, and assurance. NASA Contractor Report CR-1999-209347, NASA Langley Research Center (June 1999); also to be issued by the FAA
27. Russo, A., Hughes, J., Naumann, D.A., Sabelfeld, A.: Closing internal timing channels by transformation. In: Okada, M., Satoh, I. (eds.) ASIAN 2006. LNCS, vol. 4435, pp. 120–135. Springer, Heidelberg (2008)
28. Sabelfeld, A., Mantel, H.: Static confidentiality enforcement for distributed programs. In: Hermenegildo, M.V., Puebla, G. (eds.) SAS 2002. LNCS, vol. 2477, pp. 376–394. Springer, Heidelberg (2002)
29. Sabelfeld, A., Sands, D.: Probabilistic noninterference for multi-threaded programs. In: Proceedings of the 13th IEEE Workshop on Computer Security Foundations, CSFW 2000, p. 200. IEEE Computer Society, Washington, DC (2000)
30. Skarin, D., Barbosa, R., Karlsson, J.: Goofi-2: A tool for experimental dependability assessment. In: Proceedings of the 2010 IEEE/IFIP International Conference on Dependable Systems and Networks (2010)
31. Wahbe, R., Lucco, S., Anderson, T.E., Graham, S.L.: Efficient software-based fault isolation. In: Proceedings of the Fourteenth ACM Symposium on Operating Systems Principles, SOSP 1993, pp. 203–216. ACM, New York (1993), http://doi.acm.org/10.1145/168619.168635
32. Wang, N.J., Quek, J., Rafacz, T.M., Patel, S.J.: Characterizing the effects of transient faults on a high-performance processor pipeline. In: International Conference on Dependable Systems and Networks, DSN 2004 (2004)
33. Weber, D.G.: Formal specification of fault-tolerance and its relation to computer security. In: Proceedings of the 5th International Workshop on Software Specification and Design, IWSSD 1989, pp. 273–277. ACM, New York (1989)

Quantitative Security Analysis for Programs with Low Input and Noisy Output

Tri Minh Ngo and Marieke Huisman

University of Twente, Netherlands
tringominh@gmail.com,
Marieke.Huisman@ewi.utwente.nl

Abstract. Classical quantitative information flow analysis often considers a system as an information-theoretic channel, where private data are the only inputs and public data are the outputs. However, for systems where an attacker is able to influence the initial values of public data, these should also be considered as inputs of the channel. This paper adapts the classical view of information-theoretic channels in order to quantify information flow of programs that contain both private *and* public inputs.

Additionally, we show that our measure also can be used to reason about the case where a system operator on purpose adds noise to the output, instead of always producing the correct output. The noisy outcome is used to reduce the correlation between the output and the input, and thus to increase the remaining uncertainty. However, even though adding noise to the output enhances the security, it reduces the reliability of the program. We show how given a certain noisy output policy, the increase in security and the decrease in reliability can be quantified.

1   Introduction

Qualitative security properties, such as *noninterference* [12] and *observational determinism* [27,14], are essential for applications where private data need strict protection, such as Internet banking, e-commerce, and medical information systems, since they prohibit any information flow from a high security level to a low security level[1]. However, for many applications in which we want or need to reveal information that depends on private data, these absolutely confidential properties are not appropriate. A typical example is a *password checker* (*PWC*) where an attacker (user) tries a string to guess the password [11,3,24]. Even when the attacker makes a wrong guess, secret information has been leaked, i.e., it reveals information about what the real password is *not*. Thus, despite the correct functioning, *PWC* is rejected.

Therefore, an alternative approach for such applications is to relax the absolute properties by quantifying the information flow and determining how much

---

[1] For simplicity, throughout this paper, we consider a simple two-point security lattice, where the data is divided into two disjoint subsets, of private (high) and public (low) security levels, respectively.

J. Jürjens, F. Piessens, and N. Bielova (Eds.): ESSoS 2014, LNCS 8364, pp. 77–94, 2014.

secret information has been leaked. This information can be used to decide whether we can tolerate *minor* leakages. A quantitative security theory can be seen as a generalization of an absolute one.

Classical quantitative security analysis. Classical quantitative theory sees a program as a channel in the information-theoretic sense, where the secret S is the *only* input and the observable final outcomes O are the output [3]. An attacker, by observing O, might be able to derive information about S. The quantitative analysis of information flow then concerns the amount of private data that an attacker might learn. The analysis is based on the notion of *entropy*. The entropy of a random private variable expresses the *uncertainty* of an attacker about its value, i.e., how *difficult* it is for an attacker to discover its value. The leakage of a program is typically defined as the difference between the secret's initial uncertainty, i.e., the uncertainty of the attacker about the private data before executing the program, and the secret's remaining uncertainty, i.e., the uncertainty of the attacker after observing the program's public outcomes, i.e.,

Information leakage = Initial uncertainty - Remaining uncertainty.

Programs that contain low input. This paper considers programs where an attacker is able to influence the initial values of low variables. This is a popular kind of programs with many real-world applications, e.g., login systems, the *PWC*, or banking system. For such programs, in addition to the secret, the initial low values are another input to the channel. Therefore, the traditional form of channel becomes *invalid* for such programs.

To apply the traditional channel where the only input is the secret to this situation, we consider the initial low values as *parameters* of the channel. In particular, we consider a collection of sets of initial low values, and for each set, we construct a channel corresponding to these low values. Each channel is seen as a *test*, i.e., the attacker sets up the low parameters to test the system. Since the attacker knows the program code, then he knows which test would help him to gain the most information. Therefore, the leakage of the program with low input is defined as the maximum leakage over all possible tests.

A new measure for the remaining uncertainty. The classical approaches of the one-try attack model often base the analysis on the Smith's definition of conditional min-entropy [24]. However, the literature also admits that there might be different measures for different situations [5]. This paper argues that in some cases, Cachin's version of conditional min-entropy [7] might be a more reasonable measure, i.e., it gives more intuitive-matching results than Smith's version. Thus, we propose to consider Cachin's version as a valid measure for the remaining uncertainty. We believe that this measure has not previously been used in the theory of quantitative information flow.

The literature argues that observable outcomes would reduce the initial uncertainty of the attacker on the secret; and thus, the value of leakage cannot be negative. However, we show that this non-negativeness property does not always hold, for example in case the output of the program contains noise. The idea is that to enhance the security, the system operator might add noise to the output, i.e., instead of always producing the exact outcomes, the program might sometimes report a noisy one. The noisy-output policy makes the outcomes of program more random, and thus, it reduces the correlation between the output and the input. As a consequence, the noisy-output policy increases the remaining uncertainty, and the value of leakage might become negative. This property might open the door for a new understanding of how the measure of uncertainty should be.

To design a noisy-output policy. Adding noise enhances the security, but reduces the program's reliability, i.e., the probability that a program produces the correct outcomes. The totally random output might achieve the best confidentiality, but these outcomes are practically useless. Thus, it is clear that a noisy-output policy should consider the balance between confidentiality and reliability.

This paper discusses how to construct an efficient noisy-output policy such that the attacker cannot derive secret information from the public outcomes, while a certain level of reliability is still preserved. Since the policy is kept in secret, i.e., we do not want the attacker to find out that the system has been modified, the policy needs to satisfy some properties of the system. In this way, the noisy-output policy would help to protect the system effectively, while it still preserves the program's function at the same time.

Contributions. We propose a model of quantitative security analysis for programs that contain low input. This kind of programs has a vast application, i.e., any system with user interface. Examples of such systems include login systems, web-based applications, or online banking systems, to name a few. We also propose to consider Cachin's version of conditional min-entropy as a new measure for the notion of remaining uncertainty in the model of one-try attack. Besides, we also discuss an important property of the information flow, i.e., the quantity of information flow might be negative. This observation might change the classical view of how to define the quantity of leakage. Finally, we give an algorithm to generate noisy outcomes, while still preserving a certain level of the system's reliability. This idea can be implemented as a policy to enhance the security for applications.

Organization of the Paper. Section 2 presents the preliminaries. Then, Section 3 discusses the classical analysis, and presents our quantitative security analysis model for programs that contain low input. We also show the application of our measure. Section 4 discusses when negative information flow is expected, and how to construct and evaluate a noisy-output policy. Section 5 discusses related work, while Section 6 concludes, and discusses future work.

2 Preliminaries

2.1 Probabilistic Distribution

Let X be a discrete random variable with the carrier $\mathbf{X} = \{x_1, \ldots, x_n\}$. A *probability distribution* π over a set $\mathbf{X}$ is a function $\pi : \mathbf{X} \to [0, 1]$, such that the sum of the probabilities of all elements over $\mathbf{X}$ is 1, i.e., $\sum_{x_i \in \mathbf{X}} \pi(x_i) = 1$. If $\mathbf{X}$ is uncountable, then $\sum_{x_i \in \mathbf{X}} \pi(x_i) = 1$ implies that $\pi(x_i) > 0$ for countably many $x_i \in \mathbf{X}$. The probabilistic behavior of X is then simply given by probabilities $p(X = x_i) = \pi(x_i)$.

When X is clear from the context, we use the notation $\pi = \{p(x_1), p(x_2), \ldots, p(x_n)\}$ to denote the probabilities of elements in $\mathbf{X}$, i.e., $p(X = x_i)$.

2.2 Min-entropy

Let X and Y denote two discrete random variables. Let $p(X = x)$ denote the probability that $X = x$, and let $p(X = x | Y = y)$ denote the *conditional* probability that $X = x$ when $Y = y$.

Definition 1. *The Rényi's min-entropy of a random variable X is defined as* [24]: $\mathcal{H}_{R\acute{e}nyi}(X) = -\log \max_{x \in \mathbf{X}} p(X = x)$.

Rényi did not define the notion of conditional min-entropy, and there are different definitions of this notion.

Definition 2 (Smith's version of conditional min-entropy [24]). *The conditional min-entropy of a random variable X given Y is,*

$$\mathcal{H}_{Smith}(X | Y) = -\log \sum_{y \in \mathbf{Y}} p(Y = y) \cdot \max_{x \in \mathbf{X}} p(X = x | Y = y).$$

Definition 3 (Cachin's version of conditional min-entropy [7]). *The conditional min-entropy of a random variable X given Y is,*

$$\mathcal{H}_{Cachin}(X | Y) = -\sum_{y \in \mathbf{Y}} p(Y = y) \cdot \log \max_{x \in \mathbf{X}} p(X = x | Y = y).$$

2.3 Information-Theoretic Channel

The quantitative security analysis in the information-theoretic sense models the system as a channel with the secret as the input and the observables as the output. Formally, an information-theoretic channel is a triple $(\mathbf{X}, \mathbf{Y}, M)$, where $\mathbf{X}$ represents a finite set of secret inputs, $\mathbf{Y}$ represents a finite set of observable outputs, and M is a $|\mathbf{X}| \times |\mathbf{Y}|$ channel matrix which contains the conditional probabilities $p(y|x)$ for each $x \in \mathbf{X}$ and $y \in \mathbf{Y}$. Thus, each entry of M is a real number between 0 and 1, and each row sums to 1.

2.4 Basic Settings for the Analysis

To argue why a program is considered more dangerous than another, we need to set up some basic settings for the discussion. First, we assume that programs always terminate, and the attacker knows its source code. To aim for simplicity and clarity, rather than full generality, following [24], we restrict to programs with just a single high security input S and a single low security input L. Since the high security output is irrelevant, programs only give a low security outcome O. Our goal is to quantify how much information about S is deduced by the attacker who can influence L, and observe the execution traces of O, i.e., a sequence of values of O obtained from the program's execution. We also assume that the sets of possible values of data are finite, as in the traditional approaches.

Secondly, we assume that there is a *priori, publicly-known* probability distribution on the high values. We also assume that data at the same security level are indistinguishable in the *security meaning*. Thus, a system that leaks the last 9 bits of private data is considered to be just as dangerous as a system that leaks the first 9 bits. Finally, we consider the *one-try guessing model*, i.e., observing the public outcomes, the attacker is allowed to guess the value of S by only one try. This model of attack is suitable to many security situations where systems trigger an alarm if an attacker makes a wrong guess. For the password checker, this one-try guessing model can be understood as that an attacker is only allowed to try once. If the entered string is not the correct password, the system will block the account.

Notice that these restrictions aim to demonstrate our core idea. However, the analysis might be adapted to more complex situations easily after some trivial modifications.

3 Quantitative Security Analysis for Programs with Low Input

Before introducing our model of analysis for programs that contain low input, we present the classical models, and discuss briefly their shortcomings (see [22] for a detailed discussion).

3.1 Classical Models of Quantitative Security Analysis

Classical works [21,9,8,20,19,28,24,6] use information theory to analyze information flow quantitatively. A program is seen as a standard channel with S as the input and O as the output. Let $\mathcal{H}(S)$ denote the uncertainty of the attacker on the secret before executing the program, and $\mathcal{H}(S|O)$ the uncertainty after the program has been executed and public outcomes are observed. The leakage of the program is defined as $\mathcal{L}(P) = \mathcal{H}(S) - \mathcal{H}(S|O)$, where $\mathcal{L}(P)$ denotes the leakage of P; $\mathcal{H}$ might be either Shannon entropy or min-entropy with Smith's version of conditional min-entropy.

Classical Measures might be Counter-Intuitive. Many researchers [21,9,8,20,19,28] quantify information flow with Shannon entropy [23]. However, Smith [24] shows that in context of the one-try threat model, the Shannon-entropy measure does not always result in a very good operational security guarantee. In particular, Smith [24] shows that this measure might be counter-intuitive, i.e., an intuitively more secure program leaks more information according to the measure.

For this reason, Smith develops a new measure based on min-entropy [24]. He defines uncertainty in terms of *vulnerability* of the secret to be guessed correctly in one try. The vulnerability of a random variable X is the maximum of the probabilities of the values of X. This approach seems to match the intuitive idea of the one-try threat model, i.e., the attacker always chooses the value with the maximum probability. However, in [22], we show that the measure with Smith's version of conditional min-entropy still results in counter-intuitive values of leakage. Therefore, we agree with Alvim et al. [5]: no single leakage measure is likely to suit all cases.

Leakage in Intermediate States. Classical analysis often considers only leakages in the final states of the execution. However, for programs that contain parallel operators, the leakages in intermediate states should also be taken into account [27,14,25]. Consider the following example,

$$O := 0;$$
$$\{\text{if } (O = 1) \text{ then } (O := S) \text{ else skip}\} \parallel O := 1;$$
$$O := 1;$$

For notational convenience, let C_1 and C_2 denote the left and right operands of the parallel composition operator $\parallel$. Executing this program, we obtain the following traces $T_{|o}$ of O, depending on which thread is picked first, i.e., $T_{|o} = [0, 1, 1]$ if executing C_1 first, or $T_{|o} = [0, 1, S, 1]$ if executing C_2 first.

This program does not leak information in the final states, since the final values of O are independent of the initial value of S. However, when C_2 is executed first, the attacker is able to access S via an intermediate state.

Thus, to obtain a suitable model of quantitative security analysis, we need to consider the leakage given by a sequence of publicly observable data obtained during the execution of the program.

3.2   Leakage of Programs with Low Input

The only input of the information-theoretic channel is the secret. For programs where an attacker might influence the initial value of the low variable, the initial low value is also an input of the channel modeling the program. To use the traditional channel, we model such a program by a set of channels. Each channel corresponds to the case where the low input is assigned a specific value. Thus, in our approach, the initial low value is considered as a parameter. Since we assume that the low value set is finite, the set of channels is also finite.

We see a channel as a *test*. We run the analysis on the set of tests. Since the attacker knows the program code, and is also able to influence the initial low value, he knows which test would give him more secret information. Thus, the leakage of the program that contains low input is defined as the maximum leakages over all tests.

Given a program P that contains a low input L. Let π denote the priori distribution on the possible values of the private data, and $LVal$ denote the value set of L. Let $T_{|o}$ denote a trace of O obtained from the execution of P, i.e., the sequence of O that occurs during the execution. To define the leakage of P, we carry out the following steps.

——Leakage of Programs with Low Input——

1: Set up a test (P, π, L):

    1.1: Choose a value for L.

    1.2: Construct a channel where S is the input, L is the parameter of the channel, and the traces $T_{|o}$ are the output.

2: Compute the leakage of the test (P, π, L):

$$\mathcal{L}(P, \pi, L) = \mathcal{H}_{R\acute{e}nyi}(S) - \mathcal{H}_{Cachin}(S|L, T_{|o}),$$

    where $\mathcal{H}_{R\acute{e}nyi}(S)$ is the min-entropy of S corresponding to π.

3: Define the leakage of P as: $\mathcal{L}(P, \pi) = \max_{L \in LVal} \mathcal{L}(P, \pi, L)$.

Notice that Step **1** and **2** repeat for all values of L.

Measures of Uncertainty. Since we follow the one-try attack model, the initial uncertainty is computed as Rényi's min-entropy of S with distribution π. In this work, we propose to use Cachin's conditional min-entropy as a measure for the remaining uncertainty. Notice that in the remainder of this paper, to denote our measure, we use the notation $\mathcal{L}_{Cachin}$, instead of $\mathcal{L}$, to distinguish between our measure and Smith's measure, i.e., $\mathcal{L}_{Smith}$.

3.3   Case Studies

Below, we analyze some case studies, and compare Smith's measure with our measure. We show that our measure agrees more with the intuition.

Password Checker. Consider the following PWC. Let S denote the password, L the string entered by the attacker (low input), and O the public answer,

$$\texttt{if } (S = L) \texttt{ then } O := 1 \texttt{ else } O := 0;$$

Assume that S might be A_1, A_2, or A_3, with $\pi = \{p(A_1) = 0.98, p(A_2) = 0.01, p(A_3) = 0.01\}$. Since the attacker tests L based on the value of S, there are 3 corresponding tests, i.e., $L = A_1$, $L = A_2$, or $L = A_3$. The leakages of the tests $L = A_2$ and $L = A_3$ are the same. Hence, we only analyze $L = A_1$ and $L = A_2$.

Before interacting with the PWC, the attacker believes that the password is A_1, since $p(A_1)$ dominates the other cases. Thus, in both tests, the attacker's initial uncertainty about S is $\mathcal{H}_{R\acute{e}nyi}(S) = -\log 0.98 = 0.02915$.

When $L = A_1$, the PWC is modeled by the following channel M,

M	$O = 1$	$O = 0$
$S = A_1$	1	0
$S = A_2$	0	1
$S = A_3$	0	1

The channel M and the distribution π and determine the joint probability matrix J, where $J[s, o] = \pi(s) \cdot M[s, o]$.

J	$O = 1$	$O = 0$
$S = A_1$	0.98	0
$S = A_2$	0	0.01
$S = A_3$	0	0.01

The joint probability matrix J determines a marginal distribution of O, i.e., $p(o) = \sum_{\forall s} J[s, o]$. Thus, $p(O = 1) = 0.98$ and $p(O = 0) = 0.02$. Since $p(S = s | O = o) = \frac{J[s,o]}{p(o)}$, then $p(S = A_1 | O = 1) = 1$, $p(S = A_2 | O = 1) = p(S = A_3 | O = 1) = 0$, and $p(S = A_1 | O = 0) = 0$, $p(S = A_2 | O = 0) = p(S = A_3 | O = 0) = 0.5$. Thus, $\mathcal{L}_{Smith}(P, \pi, A_1) = 0.01465$, while $\mathcal{L}_{Cachin}(P, \pi, A_1) = 0.00915$.

When $L = A_2$, we obtain the following channel,

M	$O = 1$	$O = 0$
$S = A_1$	0	1
$S = A_2$	1	0
$S = A_3$	0	1

Thus, $p(O = 1) = 0.01$ and $p(O = 0) = 0.99$, and $p(S = A_1 | O = 1) = p(S = A_3 | O = 1) = 0$, $p(S = A_2 | O = 1) = 1$, and $p(S = A_1 | O = 0) = 0.9899$, $p(S = A_2 | O = 0) = 0$, $p(S = A_3 | O = 0) = 0.0101$. Therefore, $\mathcal{L}_{Cachin}(P, \pi, A_2) = 0.01465$, while $\mathcal{L}_{Smith}(P, \pi, A_2) = 0.01465$. The measure proposed by Smith judges that the leakages of the two tests where $L = A_1$ and $L = A_2$ are the same. However, this contradicts the intuition. In the test $L = A_1$, if the PWC answers yes, it only helps the attacker to confirm something that he already believed to be certainly true. However, if the answer is $O = 0$, it does not help the attacker at all, i.e., he still does not know whether either A_2 or A_3 is more likely to be the password, since the posteriori probability $p(S = A_2 | O = 0)$ is still equal to $p(S = A_3 | O = 0)$.

Intuitively, the test $L = A_2$ helps the attacker gain more secret information. If $O = 1$, it completely changes the attacker's priori belief, i.e., the password is not A_1, and it also confirms a very rare case, i.e., the password is A_2. If $O = 0$, this even strengthens what the attacker's belief about the secret, since the posteriori probability $p(S = A_1 | O = 0) = 0.9899$ increases. The analysis should indicate that the test $L = A_2$ leaks more information than the test $L = A_1$.

Thus, in this example, our measure gives results that match more the intuition. The leakage of this PWC is defined as the leakage of the test $L = A_2$. This example also shows that the test in which the attacker sets the low input based on the value that he believes to be the private data is not always the "best test". Since the attacker knows the source code of the program and the priori distribution of the private data, he knows which test would give him the most information. This is the reason that we define the leakage of a program with low input as the maximum leakage over all tests.

In the general case, given $\pi = \{p(A_1) = a, p(A_2) = b, p(A_3) = c\}$, whenever $a > c$ and $b > c$, Smith's measure cannot distinguish between the test $L = A_1$ and $L = A_2$, while our measure can and also agrees more with the intuition about what the leakage should be.

A Multi-threaded Program. Consider the following example,

```
O := 0;
{if (O = 1)  then   O := S/4  else   O := S mod 2} || O := 1;
O := S mod 4;
```

where S is a 3-bit unsigned integer with the priori uniform distribution. The execution of this program results in the following traces of O, depending on whether C_1 or C_2 is picked first:

S	0	1	2	3	4	5	6	7
$T\mid o$	0 0	0 0	0 0	0 0	0 0	0 0	0 0	0 0
	0 1	1 1	0 1	1 1	0 1	1 1	0 1	1 1
	1 0	1 0	1 0	1 0	1 1	1 1	1 1	1 1
	0 0	1 1	2 2	3 3	0 0	1 1	2 2	3 3

Consider a uniform scheduler, i.e., a scheduler that picks threads with equal probability. It is clear that the last command $O := S$ mod 4 always reveals the last 2 bits of S. The first bit might be leaked with probability $\frac{1}{2}$, depending on whether the scheduler picks thread C_2 first or not. Thus, with the uniform scheduler, intuitively, the real leakage of this program is 2.5 bits.

By observing the traces of O, the attacker is able to derive secret information. For example, if the trace is 0100, the attacker can derive S precisely, since this trace is produced only when $S = 0$. If the trace is 0010, the attacker can conclude that S is either 0 or 4 with the same probability, i.e., $\frac{1}{2}$. If the trace is 0111, the possible value of S is either 1 or 5, but with different probabilities, i.e., the chance that S is 5 is $\frac{2}{3}$. Therefore, $\mathcal{L}_{Cachin}(P, \pi) = 3 - (-(\frac{6}{16} \cdot \log 1 + \frac{4}{16} \cdot \log \frac{1}{2} + \frac{6}{16} \cdot \log \frac{2}{3})) = 2.53$, while $\mathcal{L}_{Smith}(P, \pi) = 3 - (-\log(\frac{6}{16} \cdot 1 + \frac{4}{16} \cdot \frac{1}{2} + \frac{6}{16} \cdot \frac{2}{3})) = 2.58$.

Consider a scheduler that picks thread C_2 first with probability $\frac{3}{4}$. With this scheduler, the real leakage of this program is 2.75. Our measure gives $\mathcal{L}_{Cachin}(P, \pi) = 2.774$, while $\mathcal{L}_{Smith}(P, \pi) = 2.807$. If the scheduler picks thread C_2 first with probability $\frac{1}{4}$, $\mathcal{L}_{Cachin}(P, \pi) = 2.271$, while $\mathcal{L}_{Smith}(P, \pi) = 2.321$. Of course in this case, the real leakage is 2.25. These results show that our measures are closer to the real leakage values.

These case studies show that our measure is more precise than the classical measure given by Smith's conditional min-entropy. The main difference between the two measures is the position of log in the expression of the remaining entropy. The idea of using logarithm is to express the notion of uncertainty in bits. Thus, the log should apply only to the probability of the guess, which represents the uncertainty of the attacker, as in our approach. Our measure distinguishes between the probabilities of the observable and the probabilities of the guess based on the observable. In Smith's measure, the logarithm applies to the combination of the two probabilities, and does not distinguish between them, which might cause imprecise results.

However, as a side remark, we emphasize that no unique measure is likely to be suitable for all cases. We believe that for some examples, measures based on Shannon entropy or Smith's version of conditional min-entropy might match better the real values of leakage.

4   Adding Noise to the Output

4.1   Negative Information Flow

In relation to defining an appropriate measure for information flow quantification, this paper also discusses a claim of the existing theory of quantitative information flow, i.e., a quantitative measure of information leakage should return a non-negative value. The common idea of the classical analysis is that the observation of the program's public outcomes would enhance the attacker's knowledge about the private data, and consequently reduce the attacker's initial uncertainty.

However, we think that this non-negativeness property does not always hold. For some applications, to enhance the confidentiality, the system operator adds noise secretly to the output, i.e., via some output perturbation mechanism based on randomization. The noisy outcomes might mislead the attacker's belief about the secret, i.e., they increase the final uncertainty. As a consequence, the value of leakage might be negative. This idea is illustrated more as follows.

Password checker with noisy outcomes. Consider the *PWC* in Section 3.3. We assume that the system operator *secretly* has changed its behavior, i.e., the real *PWC* is a probabilistic *PWC* where the system operator introduced some perturbation mechanism to the output (We assume that the attacker does not know about the security policy applied to the system.),

$$\text{if } (S = L) \text{ then } \{O := 1 \;\;_{0.9}[\!] \; O := 0\} \text{ else } \{O := 0 \;\;_{0.9}[\!] \; O := 1\};$$

In this version, the exact answers are reported with probability 0.9, i.e., when $S = L$, $O = 1$ is reported with probability 0.9, and $O = 0$ with probability 0.1. Consider the test $L = A_2$, the real channel M' is as follows,

M'	$O = 1$	$O = 0$
$S = A_1$	0.1	0.9
$S = A_2$	0.9	0.1
$S = A_3$	0.1	0.9

Notice that the attacker still thinks that the system is M, but in fact, the real system is M'. Based on π and M', the computation gives the real distribution $p(O = 1) = 0.108$ and $p(O = 0) = 0.892$, and the real posteriori probabilities $p(S = A_2 | O = 1) = 0.083$ and $p(S = A_1 | O = 0) = 0.9887$.

Before observing the outcome, the guess, i.e., the secret is A_1, has 98% chance of being correct. If the outcome is $O = 0$, the real posteriori probability gives the attacker's guess, i.e., $S = A_1$, a 98.87% chance of being correct. This is almost the same as the guess without the outcome. When $O = 1$, the attackers guesses $S = A_2$, since his database tells that this guess has the highest chance to be correct. However, the real posteriori probability ensures that his guess only has a 8.3% chance of being correct. Therefore, the outcomes of the program not only reveal *no* secret information, but also cause him to decide wrongly. Therefore, intuitively, this is a negative information flow.

As we expected, our measure indicates a negative leakage: $\mathcal{L}_{Cachin}(P, \pi, A_2) = -\log 0.98 + (0.108 \log 0.083 + 0.892 \log 0.9887) = -0.37$, while $\mathcal{L}_{Smith}(P, \pi, A_2) = -0.137$. Notice that the value of the leakage is determined by the real probability of success, not by the probability in the attacker's database.

We believe that this observation of negative information flow has not been reported in the literature. We think that this property would change the classical view of how the measure of uncertainty should be, i.e., we do not need to avoid measures that do not guarantee the non-negativeness property.

4.2 Noisy-Output Policy

The noisy outcomes change the behavior of the system, i.e., they change the channel M that models the system (the public channel that the attacker also knows) to M' (the real channel in secret). The noisy outcomes should be added in such a way that they change the original channel, but still preserve a certain level of reliability, e.g., the above probabilistic *PWC* works properly in 90% of the time. Totally random outcomes might achieve the best confidentiality, but these outcomes are practically useless. Besides, the noisy-output policy also needs to satisfy some *general* requirements that, on one hand, help to mislead the attacker, i.e., the attacker does not know that the system has been changed by the policy; thus, he still uses the posteriori distributions based on M and π to make a guess, and on the other hand, reduce the leakage. This section discusses how to design such an efficient noisy-output policy.

Design a Policy. Given a system P that is described by a channel matrix M of size $n \times m$, e.g., the set of secret input values is $\{A_1, \cdots, A_n\}$, and the set of observable outcomes is $\{Z_1, \cdots, Z_m\}$.

General Requirements. Since the attacker knows π and M, he is able to compute the marginal distribution of the output. Thus, firstly, the *distribution of the output* has to be preserved by the channel M', where the noise has been added. If the policy does not preserve this distribution, the attacker might find out that the channel M has been changed, and then he will try to study the system before making a guess, i.e., trying to get the *real* program code of the system.

Secondly, for each outcome Z_i, assume that $p(S = A_j|Z_i)$ is the maximum posteriori probability, then $p'(S = A_j|Z_i)$ is also the maximum posteriori probability, i.e., the *maximum property* of the posteriori distributions has to be preserved.

For example, if M gives a posteriori distribution where $p(S = A_j|Z_i) = 0.8$, then the real posteriori probability given by M' might be $p(S = A_j|Z_i) = 0.6$. Thus, if the outcome is Z_i, the attacker thinks that the guess $S = A_j$ has a 80% chance to be correct. However, in reality, this guess only has a 60% chance of success. Notice that $p(S = A_j|Z_i)$ does not need to be equal to $p'(S = A_j|Z_i)$. The preservation of the maximum property of the posteriori distribution is necessary. Consider a uniform posteriori distribution $\{p(S = A_1|Z_i) = p(S = A_2|Z_i) = p(S = A_3|Z_i) = \frac{1}{3}\}$ in the attacker's database. Following the requirement, the posteriori distribution given by M' has to be also uniform. If we do not require this, then the real distribution might possibly be $\{p'(S = A_1|Z_i) = 0.2, p'(S = A_2|Z_i) = 0.7, p'(S = A_3|Z_i) = 0.1\}$. According to his database, the attacker might guess $S = A_2$, since all three guesses have the same chance of being correct. In this case, the real probability would increase the chance of success, and thus, increase the leakage.

Reliability. Reliability of a system is the probability that a system will perform its intended function during a specified period of observation time. Let $\mathcal{R}_i$ ($\mathcal{R}_i \leq 1$) denote the reliability corresponding to the secret value A_i, i.e., the probability that the system will produce correct outcomes when the secret is A_i. Thus, the overall reliability of the system P is $\mathcal{R}_P = \sum_i p(A_i) \cdot \mathcal{R}_i$. The noisy-output policy produces noise, and thus it reduces the reliability of the system. Therefore, we require that a noisy-output should guarantee at least a certain level of reliability.

Noisy-output policy. We propose a simple policy that might reduce the unwanted information flow, while still preserving a certain level of reliability. The following policy only aims to demonstrate the core idea of what a noisy-output policy should be. The practical policy might be customized due to the requirement of the application. Given a channel M that models a system P. A noisy-output policy changes M to M' by choosing an appropriate set of $\{\mathcal{R}_1, \cdots, \mathcal{R}_n\}$.

────Noisy-Output Policy─────────────────────────────

1: For each row i of M, multiply each entry of the row by the reliability *variable* $\mathcal{R}_i$. Choose randomly one of the smallest entries, and add the value $1 - \mathcal{R}_i$ to it. Denote this modified matrix by M'.

2: Choose an overall reliability value that the policy has to guarantee, e.g., $\mathcal{R}_{min}$. Establish an inequality: $\sum_i p(A_i) \cdot \mathcal{R}_i \geq \mathcal{R}_{min}$.

3: For any outcome Z_i, let $p(O = Z_i)$ denote the probability determined by π and M, and $p'(O = Z_i)$ determined by π and M', establish an equation: $p(O = Z_i) = p'(O = Z_i)$.

4: For each outcome Z_i, if $\forall k. p(S = A_j|Z_i) \geq p(S = A_k|Z_i)$, then establish the following condition: $\forall k. p'(S = A_j|Z_i) \geq p'(S = A_k|Z_i)$.

5: Solve these equations and inequalities. The set $\{\mathcal{R}_1, \cdots, \mathcal{R}_n\}$, are chosen in such a way that the leakage given by M' is close to zero, and the reliability of the system $\mathcal{R}_P$ is as high as possible.

──

Notice that in Step 1, the sum of all entries of a row has to be 1; thus we have to add the value $1 - \mathcal{R}_i$ to one of its entries. Step 3 establishes a set of equations, i.e., $m - 1$ *independent* equations that correspond to $m - 1$ observable outcomes, that preserve the output distribution. We also obtain $(n - 1) \cdot m$ inequalities in Step 4, that preserve the maximum property of the posteriori distributions.

There always exists a trivial solution $\mathcal{R}_1 = \cdots = \mathcal{R}_n = 1$, i.e., M and M' are identical. When there are multiple solutions, we choose one that gives a low leakage, but a high overall reliability. However, this does not always happen. A solution that guarantees a very low leakage might also give a low reliability. In fact, a negative leakage, i.e., when the attacker decides wrongly based on the observable outcomes, is not always necessary. The goal of the policy is to ensure that the attacker cannot gain knowledge from the observable outcomes. Thus, $\mathcal{R}_1, \cdots, \mathcal{R}_n$ are chosen such that the leakage is close to zero and the overall reliability gets a high value. Next, we show an important property of our policy.

Theorem 1. *Given a priori distribution π and a channel matrix M, the channel matrix M' modified from M by the noisy-output policy always gives a leakage quantity that is not greater than the one given by M.*

Proof. For any outcome Z_i, assume that the maximum likely secret is A_j. Since $p'(Z_i) = p(Z_i)$ and $p'(S = A_j|Z_i) = \mathcal{R}_j \cdot p(S = A_j|Z_i)$, thus $-p'(Z_i) \log p'(S = A_j|Z_i) \geq -p(Z_i) \log p(S = A_j|Z_i)$. Therefore, the value of remaining uncertainty given by π and M' is greater than or equal to the one given by π and M. As a consequence, the corresponding leakage quantity is reduced.

Example. Consider a deterministic program P where the secret might be A_1, A_2, or A_3 with a uniform $\pi = \{p(A_1) = p(A_2) = p(A_3) = \frac{1}{3}\}$. The system P might produce three low outcomes Z_1, Z_2, and Z_3 as described by M,

M	Z_1	Z_2	Z_3
$S = A_1$	1	0	0
$S = A_2$	0	1	0
$S = A_3$	0	0	1

Since the attacker knows the program code, he is able to construct M in his database. Since the public outcomes are totally dependent on the secret, the attacker can derive the private data entirely from the outcomes, e.g., if the outcome is Z_i, the attacker knows for sure that $S = A_i$.

To protect the secret, the system operator might mislead the attacker by adding noise to the output, i.e., the real system is M',

M'	Z_1	Z_2	Z_3
$S = A_1$	$\mathcal{R}_1$	$1 - \mathcal{R}_1$	0
$S = A_2$	0	$\mathcal{R}_2$	$1 - \mathcal{R}_2$
$S = A_3$	$1 - \mathcal{R}_3$	0	$\mathcal{R}_3$

Based on π and M, the attacker knows that $p(O = Z_1) = p(O = Z_2) = p(O = Z_3) = \frac{1}{3}$. To satisfy this output distribution, $\mathcal{R}_1 = \mathcal{R}_2 = \mathcal{R}_3$. Besides, the maximum property of the posteriori distributions determines that

$\frac{1}{2} \leq \mathcal{R}_1, \mathcal{R}_2, \mathcal{R}_2 \leq 1$. Thus, the reliability of the system is $\mathcal{R}_P = \mathcal{R}_1$, and for this example, $\mathcal{L}_{Cachin}(P, \pi) = \mathcal{L}_{Smith}(P, \pi) = \log 3\mathcal{R}_1$.

Thus, a high value of $\mathcal{R}_1$, which guarantees a high overall reliability, also gives a high leakage. If the goal is to reduce the leakage, we might choose $\mathcal{R}_1 = \mathcal{R}_2 = \mathcal{R}_3 = \frac{1}{2}$, which gives the smallest value of leakage, i.e., $\log \frac{3}{2}$, but also a very low reliability. If a high reliability is required, $\mathcal{R}_1 = \mathcal{R}_2 = \mathcal{R}_3 = \frac{2}{3}$ might be a good choice.

Consider the PWC example, for the test $L = A_2$, following the policy, we can choose $\mathcal{R}_1 = 0.995, \mathcal{R}_2 = 0.5, \mathcal{R}_3 = 0.99$ to have $\mathcal{L}_{Cachin}(P, \pi, A_2) = -0.00275$ with the reliability $\mathcal{R} = 0.99$. However, if we consider both tests, i.e., $L = A_1$ and $L = A_2$, $\mathcal{R}_1 = \mathcal{R}_2 = \mathcal{R}_3 = 1$.

As mentioned above, a noisy-output policy enhances the security, but reduces the reliability of a system, i.e., the system does not always work in a proper way. However, the drawback of the reduced reliability can be overcome. Consider a situation of the PWC in which an user or an attacker provides a *correct* password, but the system rejects it, and then blocks his account (the one-try model). If this context is for the attacker, it would be very nice, since the attacker does not have a chance to use the account again. If this context is for the real user; however, the situation would be different: the user is still allowed to reactivate the account by contacting the company/website administrators and proving that he is the real owner of the account, while the attacker cannot do the same.

The other way around, i.e., when the system accepts a wrong password, is not nice for the security. This is the reason that the policy should guarantee a high reliability. Notice that in this situation, the system accepts the login, but no private information is leaked, since the attacker still does not know the correct password. Thus, in the next login, there is a high chance that the system will reject this wrong password. Moreover, to avoid this situation, the system might also implement two-factor authentication, i.e., in addition to asking for something that only the user knows (e.g., user-name, password, PIN), the system also requires something that only the user has (e.g., ATM card, smart card). The ATM scenario illustrates the basic concept of most two-factor authentication systems, i.e., without the combination of both ATM card and PIN verification, authentication does not succeed.

Finally, it would be stressed that we only sketch the main idea of a noisy-output policy. However, for practical applications, depending on the real security requirements, the above policy might be customized, e.g., in Step **1**, instead of choosing randomly one of the smallest entries, and adding $1 - \mathcal{R}_i$ to it, the policy can add to each of the smallest entries a value such that the sum of all these values is equal to $1 - \mathcal{R}_i$.

5   Related Work

Our proposal for programs with low input borrows ideas from Malacaria et al. [20] and Yasuoka et al. [26]. However, in these works, they [20,26] do not analyze the systems with low input sufficiently. They define the leakage of a program as the leakage of a single test, while we define it as the maximum leakage of all tests. All examples in [20] are without low input, and their measure is based on Shannon entropy. Yasuoka et al. only consider the leakages in the final states.

Clarkson et al. [11] argue that the classical uncertainty-based analysis is not adequate to measure information flow of programs that contain low input. To define the leakage of the *PWC*, Clarkson et al. fix the value of the secret, i.e., the password, and the low input, i.e., by always assigning it the value that the attacker believes to be the password, then run the analysis under that specific circumstance. In case the value the attacker believes is not the real password, the uncertainty-based analysis might return a negative leakage value. The authors argue that this result flatly contradicts the intuition, i.e., from interacting with *PWC*, the attacker gains more knowledge by learning that the password is not the one that he has just entered. Based on this claim, they propose a different approach named accuracy-based information flow analysis. This trend of research has been expanded in [10,15,16,13]. However, the accuracy-based analysis often results in a quantity that is *inconsistent* with the size of the flow, i.e., the quantity of the secret information flow exceeds the size needed to store the secret [10].

We believe that there is a flaw in the way Clarkson et al. model the system. Clarkson et al. fix the value of the secret. Thus, this does not capture precisely the idea of information-theoretic channel. The information-theoretic channel has the secret as the input, and the entropy of the input quantifies the uncertainty involved in predicting the value of the secret. Thus, if the value of the secret is fixed, it implies that the priori distribution on the possible values of the secret is not valid anymore, i.e., the secret is now a certain value with the absolute probability 1. As a consequence, the entropy of the input does not reflect the true meaning of the initial uncertainty. Therefore, in these approaches, a wrong channel model has led to misleading results, i.e., a negative uncertainty-based result, or a size-inconsistent accuracy-based result.

Alvim et al. discuss limitations of the classical information-theoretic channel, i.e., showing that it is not a valid model for interactive systems where secrets and observables can alternate during the computation and influence each other [4]. In [3], Alvim et al. also discuss the example of the password checker. They fix the password by assigning it a specific value, and then consider the initial low values as the only input to the channel. As discussed before, this idea does not reflect the true idea of the information-theoretic channel.

Köpf et al. also consider systems with low input, i.e., cryptosystems where the attacker can control the set of input messages [17]. However, their proposal is only for deterministic systems, i.e., for each input, the system produces only one output, while in our proposal, the output might be nondeterministic and probabilistic. Besides, Köpf et al. consider a different threat model, i.e., the *multiple-try* guessing model, and they put a restriction on the priori distribution of the secret, requiring it to be uniform.

The idea of adding noise to the output comes from the *differential privacy* control, i.e., the problem of protecting the privacy of database's participants when performing *statistical* queries [3,1,2]. The differential privacy control also uses some output perturbation mechanism to report a noisy answer among the correct ones for the queries. Thus, while the attacker is still able to learn properties of the population as a whole, he cannot learn the value of an individual. To construct an efficient noisy-output policy for a statistical database, it is necessary to consider the balance between *privacy*, i.e., how difficult to guess the value of an individual, and *utility*, i.e., the capacity to retrieve accurate answers from the reported ones. In [18], Köpf et al. also explore a similar idea to cope with timing attacks for cryptosystems, i.e., randomizing each cipher-text before decryption. As a consequence, the strength of the security guarantee is enhanced, while the efficiency of the cryptosystem is decreased, since the execution time of the cryptographic algorithm is increased.

In this work, we assume that the attacker cannot choose schedulers. The idea is to make our measure valid for both sequential and multi-threaded programs. Since sequential programs contain no parallel operator, the scheduler is not necessary for such programs. In [22], we propose a model of analysis for multi-threaded programs where the attacker is able to select an appropriate scheduler to control the set of program traces. In this current work, if the attacker can choose schedulers, i.e., if observations in our channel are not only traces, but also include scheduling decisions at each step, our measure and the proposed measure in [22] coincide. Notice that we did not consider the low input in [22].

6   Conclusions and Future Work

This paper discusses how to analyze quantitatively information flow of a program that contains low input. For such programs, we adapt the traditional information-theoretic channel by considering the initial low values as parameters of the channel. Besides, we also show that the value of information flow might be negative in case the system operator adds noise to the outcomes, i.e., the noise misleads the attacker's belief about the secret, and thus, it increases the final uncertainty. We believe that this property would change the way people often think about the measure of uncertainty. Since there is a growing appreciation that no unique measure is suitable for all cases, we suggest to measure the remaining uncertainty by Cachin's conditional min-entropy. This new measure matches the intuition in many cases. Finally, this paper discusses how to design an efficient noisy-output policy, which generates noisy outcomes, while still guarantees a high overall reliability.

Future Work. The classical approaches of the one-try attack model only base the analysis on the information of the value that the attacker believes to be the secret. Thus, the analysis ignores the extra leakage that might be derived from the values that the attacker disbelieves to be the secret. In the future work, we propose to include this extra information to the analysis. We also consider to define a measure for the *multiple-try* attack model.

Since there are many measures proposed for quantitative information flow analysis, and no unique measure is likely to suit all contexts, it might be interesting to evaluate each measure to determine under which circumstances, a certain measure might give the *best* answer.

Acknowledgments. The authors would like to thank Catuscia Palamidessi and Kostas Chatzikokolakis for many fruitful discussions. Our work is supported by NWO as part of the SlaLoM project.

References

1. Alvim, M.S., Andrés, M.E., Chatzikokolakis, K., Palamidessi, C.: On the relation between differential privacy and quantitative information flow. In: Aceto, L., Henzinger, M., Sgall, J. (eds.) ICALP 2011, Part II. LNCS, vol. 6756, pp. 60–76. Springer, Heidelberg (2011)
2. Alvim, M.S., Andrés, M.E., Chatzikokolakis, K., Degano, P., Palamidessi, C.: Differential privacy: on the trade-off between utility and information leakage. CoRR, abs/1103.5188 (2011)
3. Alvim, M.S., Andrés, M.E., Chatzikokolakis, K., Palamidessi, C.: Quantitative information flow and applications to differential privacy. In: Aldini, A., Gorrieri, R. (eds.) FOSAD VI 2011. LNCS, vol. 6858, pp. 211–230. Springer, Heidelberg (2011)
4. Alvim, M.S., Andrés, M.E., Palamidessi, C.: Information flow in interactive systems. In: Gastin, P., Laroussinie, F. (eds.) CONCUR 2010. LNCS, vol. 6269, pp. 102–116. Springer, Heidelberg (2010)
5. Alvim, M.S., Chatzikokolakis, K., Palamidessi, C., Smith, G.: Measuring information leakage using generalized gain functions. In: Proceedings of the IEEE 25th Computer Security Foundations Symposium, CSF 2012, pp. 265–279. IEEE Computer Society (2012)
6. Andres, M.E., Palamidessi, C., Rossum, P., Sokolova, A.: Information hiding in probabilistic concurrent systems. In: Proceedings of the 2010 Seventh International Conference on the Quantitative Evaluation of Systems, QEST 2010, pp. 17–26. IEEE Computer Society (2010)
7. Cachin, C.: Entropy Measures and Unconditional Security in Cryptography. PhD thesis (1997)
8. Chatzikokolakis, K., Palamidessi, C., Panangaden, P.: Anonymity protocols as noisy channels. In: Montanari, U., Sannella, D., Bruni, R. (eds.) TGC 2006. LNCS, vol. 4661, pp. 281–300. Springer, Heidelberg (2007)
9. Clark, D., Hunt, S., Malacaria, P.: Quantitative information flow, relations and polymorphic types. J. Log. and Comput. 15, 181–199 (2005)
10. Clarkson, M.R., Myers, A.C., Schneider, F.B.: Quantifying information flow with beliefs. J. Comput. Secur. (2009)
11. Clarkson, M.R., Myers, A.C., Schneider, F.B.: Belief in information flow. In: In Proc. 18th IEEE Computer Security Foundations Workshop, pp. 31–45 (2005)
12. Goguen, J.A., Meseguer, J.: Security policies and security models. In: IEEE Symposium on Security and Privacy, pp. 11–20 (1982)
13. Hamadou, S., Sassone, V., Palamidessi, C.: Reconciling belief and vulnerability in information flow. In: Proceedings of the 2010 IEEE Symposium on Security and Privacy, SP 2010, pp. 79–92. IEEE Computer Society (2010)

14. Huisman, M., Ngo, T.M.: Scheduler-specific confidentiality for multi-threaded programs and its logic-based verification. In: Beckert, B., Damiani, F., Gurov, D. (eds.) FoVeOOS 2011. LNCS, vol. 7421, pp. 178–195. Springer, Heidelberg (2012)
15. Hussein, S.H.: A precise information flow measure from imprecise probabilities. In: Proceedings of the 2012 IEEE Sixth International Conference on Software Security and Reliability, SERE 2012, pp. 128–137. IEEE Computer Society (2012)
16. Hussein, S.H.: Refining a quantitative information flow metric. CoRR, abs/1206.0886 (2012)
17. Köpf, B., Basin, D.: An information-theoretic model for adaptive side-channel attacks. In: Proceedings of the 14th ACM Conference on Computer and Communications Security, CCS 2007, pp. 286–296. ACM (2007)
18. Köpf, B., Dürmuth, M.: A provably secure and efficient countermeasure against timing attacks. In: Proceedings of the 2009 22nd IEEE Computer Security Foundations Symposium, CSF 2009, pp. 324–335. IEEE Computer Society (2009)
19. Malacaria, P.: Risk assessment of security threats for looping constructs. J. Comput. Secur. 18, 191–228 (2010)
20. Malacaria, P., Chen, H.: Lagrange multipliers and maximum information leakage in different observational models. In: Proceedings of the Third ACM SIGPLAN Workshop on Programming Languages and Analysis for Security, PLAS 2008, pp. 135–146. ACM (2008)
21. Moskowitz, I.S., Newman, R.E., Crepeau, D.P., Miller, A.R.: Covert channels and anonymizing networks. In: Proceedings of the 2003 ACM Workshop on Privacy in the Electronic Society, WPES 2003, pp. 79–88. ACM (2003)
22. Ngo, T.M., Huisman, M.: Quantitative security analysis for multi-threaded programs. CoRR, abs/1306.2693 (2013)
23. Shannon, C.E., Weaver, W.: A Mathematical Theory of Communication. University of Illinois Press (1963)
24. Smith, G.: On the foundations of quantitative information flow. In: de Alfaro, L. (ed.) FOSSACS 2009. LNCS, vol. 5504, pp. 288–302. Springer, Heidelberg (2009)
25. Volpano, D., Smith, G.: Probabilistic noninterference in a concurrent language. J. Comput. Secur. 7, 231–253 (1999)
26. Yasuoka, H., Terauchi, T.: On bounding problems of quantitative information flow. In: Gritzalis, D., Preneel, B., Theoharidou, M. (eds.) ESORICS 2010. LNCS, vol. 6345, pp. 357–372. Springer, Heidelberg (2010)
27. Zdancewic, S., Myers, A.C.: Observational determinism for concurrent program security. In: Proceedings of 16th IEEE Computer Security Foundations Workshop, CSFW 2003, pp. 29–43. IEEE Computer Society (2000)
28. Zhu, Y., Bettati, R.: Anonymity vs. information leakage in anonymity systems. In: Proceedings of the 25th IEEE International Conference on Distributed Computing Systems, ICDCS 2005, pp. 514–524. IEEE Computer Society (2005)

A Modeling and Formal Approach for the Precise Specification of Security Patterns

Brahim Hamid and Christian Percebois

IRIT, University of Toulouse
118 Route de Narbonne, 31062 Toulouse Cedex 9, France
{hamid,percebois}@irit.fr

Abstract. Non-functional requirements such as Security and Dependability (S &D) become more important as well as more difficult to achieve. In fact, the integration of security features requires the availability of both application domain specific knowledge and security expertise at the same time. Hence, capturing and providing this expertise by the way of *security patterns* can support the integration of S&D features by design to foster reuse during the process of software system development.

The solution envisaged here is based on combining metamodeling techniques and formal methods to represent security pattern at two levels of abstraction fostering reuse during the process of pattern development and during the process of pattern-based development. The contribution of this work is twofold: (1) An improvement of our previous pattern modeling language for representing security pattern in the form of a subsystem providing appropriate interfaces and targeting security properties, (2) Formal specification and validation of pattern properties, using the interactive Isabelle/HOL proof assistant. The resulting validation artifacts may mainly complete the definitions, and provide semantics for the interfaces and the properties in the context of S&D. As a result, validated patterns will be used as bricks to build applications through a Model-Driven engineering approach.

Keywords: Pattern, Metamodel, Domain, Formalization, Model-Driven engineering, Security.

1 Introduction

Recent times have seen a paradigm shift in terms of design by combining multiple software engineering paradigms, namely, Model-Driven Engineering (MDE) [20] and formal methods [25]. Such a paradigm shift is changing the way systems are developed nowadays, reducing development time significantly. Embedded systems [26] are a case where a range of development approaches have been proposed. The most popular are those using models as main artifacts to be constructed and maintained. In these approaches, software development consists of model specification and transformations. In addition to MDE, pattern-based development has gained more attention recently in software engineering by addressing new challenges that were not targeted in the past. In fact, they are applied in modern software architecture for distributed systems including middlewares, real-time embedded systems, and recently in security and dependability engineering.

J. Jürjens, F. Piessens, and N. Bielova (Eds.): ESSoS 2014, LNCS 8364, pp. 95–112, 2014.
© Springer International Publishing Switzerland 2014

Unfortunately, most of security patterns are expressed as informal indications on how to solve some security problems, using template like traditional patterns [1,6]. These patterns do not include sufficient semantic descriptions, including those of security and dependability concepts, for automated processing within a tool-supported development and to extend their use. Furthermore, due to manual pattern implementation, the problem of incorrect implementation (the most important source of security problems) remains unsolved. For that, model driven software engineering can provide a solid basis for formulating design patterns that can incorporate security aspects and for offering these patterns at several levels of abstraction.

In this paper, we leverage on this idea to propose a new framework for the specification and the validation of security patterns intended for systems with stringent security requirements. Reaching this target requires to get a common representation of such a modeling environment for several domains and the ability to customize them for a specific domain. Here we propose two additional and complementary types of representations: a semi-formal representation through metamodeling techniques and a rigorous formal representation through interactive theorem proving approaches. Regarding the comparison with documentation template [6] (informal representation) used to represent security patterns, our proposition is based on it. However, we keep the template elements in the form of attributes and we deeply refine them by the definition of new concepts in order to fit with security engineering needs. In this vision, a security or (in general an S&D) pattern is a subsystem exposing pattern functionalities through interfaces and targeting security properties. As it follows the MDE paradigm for system's design, using patterns on different levels of abstraction, it allows for integration into the system's design process, hence supports this process. To this end, the proposed representation takes into account the simplification and the enhancement of such activities, namely: selection/search, based on the classified properties, and integration, based on a high level description of interfaces.

The rest of this paper is organized as follows. An overview of the modeling approach we proposed including a set of definitions is presented in Section 2. Then, Section 3 presents the pattern modeling language and illustrates the pattern modeling process in practice. Section 4 presents the validation process through the example of the secure communication pattern. In Section 5, we review most related works addressing pattern specification and validation. Finally, Section 6 concludes this paper with a short discussion on future works. An appendix is added with the Isabelle/HOL code of our experiment.

2   The Nature of Patterns within PBSE (Pattern-Based System and software Engineering)

Usually, pattern design artifacts are provided as a library of models (subsystems) and as a system of patterns (framework) in the more elaborated approaches. However, there are still lacks in existing modeling languages and/or formalisms dedicated to model these design artifacts and the way how to reuse them in software development automation. The supporting research activities examine three distinct challenges: mining - discovering patterns from existing systems, hatching - selection of the appropriate pattern and application - effective use during the system development process.

In our work, we study only the two last challenges, targeting the (i) development of an extensible design language for modeling patterns for secure and dependable distributed embedded systems and (ii) a methodology to improve existing development processes using patterns. The language has to capture the core elements of the pattern to help its (a) precise specification, (b) selection and (c) integration. Supporting research tackles the presented challenges includes domain patterns, pattern languages and recently formalisms and modeling languages to foster their application in practice. Here, we propose a novel approach to improve the security pattern representation by the combination of semi-formal semantics and rigorous formal semantics of some of its concepts.

2.1 Motivational Example: Secure Communication Pattern (SCP)

As example of a common and a widely used patterns we choose the Secure Communication Pattern referred to in the following as SCP. Messages passing across any public network can be intercepted. The problem is how to ensure that data being across this system is secure in transit. In other words, how to guarantee data authenticity. This is one of the goal to use the SCP.

However, those SCP are slightly different with regard to the application domain. For instance, a system domain has its own mechanisms and means to serve the implementation of this pattern using a set of protocols ranging from SSL, TLS, Kerberos, IPSec, SSH, WS-Security and so on. In summary, they are similar in the goal, but different in the implementation issues. So, the motivation is to handle the modeling of security patterns by following abstraction. In the following, we propose to use SSL mechanisms [1] to specialize the implementation of the SCP.

The establishment of a secure channel is composed of two phases. First, the client informs the server of the cryptographic algorithms it can handle. The actual choice is always made by the server, which reports its choice back to client. In the second phase, authentication takes place. The server is required to authenticate itself. It sends a certificate containing its public key signed by a certification authority CA to the client. In order to authenticate the client by the server, the client has to send a certificate to the server as well. The client generates a random number that will be used by both sides for building a session key, and sends this number to the server, encrypted with the server's public key. In the case of client required authentication, the client signs the number with its private key. At that point, the sever can verify the identity of the client, after which the secure channel can be established.

2.2 Definitions and Concepts

In [6], a design pattern abstracts the key artifacts of a common design structure that make it useful for creating a reusable object-oriented design. Several generalizations on this basis to describe software design patterns in general are proposed in literature. We quote the following definition of security patterns from [21]:

---

[1] SSL or its update named TLS proposed in RFC 2246.

Definition 1 (Security Pattern). *A security pattern describes a particular recurring security problem that arises in specific contexts and presents a well-proven generic scheme for its solution.*

Security patterns are not only defined from a platform independent viewpoint (i.e. they are independent from the implementation), they are also expressed in a consistent way with domain specific models. Consequently, they will be much easier to understand and validate by application designers in a specific area. To capture this vision, we introduced the concept of *domain view*. Particularly a security pattern at domain independent level exhibits an abstract solution without specific knowledge on how the solution is implemented with regard to the application domain.

Definition 2 (Domain). *A domain is a field or a scope of knowledge or activity that is characterized by the concerns, methods, mechanisms, ... employed in the development of a system. The actual clustering into domains depends on the given group/community implementing the target methodology.*

In our context, a domain represents all the knowledge including protocols, processes, methods, techniques, practices, OS, HW systems, measurement and certification related to the specific domain. With regard to the artifacts used in the system under development, we will identify the first classes of the domain to specialize such artifacts. For instance, the specification of a pattern at domain independent point of view is based on the software design constructs. The specialization of such a pattern for a domain uses a domain protocol to implement the pattern solution (see example of secure communication pattern given in Section 3.2.1 and Section 3.2.2).

The objective is to reuse the domain independent model security patterns for several industrial application domain sectors and also let them be able to customize those domain independent patterns with their domain knowledge and/or requirements to produce their own domain specific artifacts. Thus, the 'how' to support these concepts should be captured in the specification languages.

3 Pattern Modeling Process

We now present an overview of our pattern specification process. Along this description, we will give the main keys to understand why our process is based on a general and a constructive approach. We begin with our pattern specification metamodel.

3.1 Pattern Specification Metamodel (SEPM)

To foster reuse of patterns in the development of critical systems with S&D requirements, we are building on a metamodel for representing security pattern in the form of a subsystem providing appropriate interfaces and targeting security properties to enforce the S&D system requirements. Interfaces will be used to exhibit pattern's functionality in order to manage its application. In addition, interfaces support interactions with security primitives and protocols for the specialization for a specific application domain. The principal classes of the System and Software Engineering Pattern Metamodel (SEPM) [11] are described with Ecore notations in Figure 1. Their meanings are more detailed in the following paragraphs.

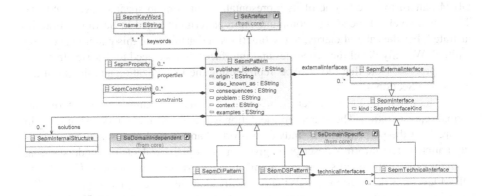

Fig. 1. The SEPM metamodel - Overview

- *SepmPattern*. This block represents a security pattern as a subsystem describing a solution for a security particular recurring design problem that arises in specific design context. A *SepmPattern* defines its behavior in terms of provided and required interfaces. Larger pieces of a system's functionality may be assembled by reusing patterns as parts in an encompassing pattern or assembly of patterns, and wiring together their required and provided interfaces. A *SepmPattern* may be manifested by one or more artifacts.
- *SepmDIPattern*. It is an *SepmPattern* denoting some abstract representation of a security pattern at domain independent level. This is the key entry artifact to model patterns at domain independent level (DIPM).
- *Interface*. A *SepmPattern* interacts with its environment with *Interfaces* which are composed of *Operations*. A *SepmPattern* owns provided and required interfaces. A provided interface highlights the services exposed to the environment. A required interface corresponds to services needed by the pattern to work properly. We consider two kinds of interface:
  - *External interface*. Allows implementing interaction with regard to the integration of a pattern into an application model or to compose patterns.
  - *Technical interface*. Allows implementing interaction with security primitives and protocols, such as encryption, and specialization for specific underlying software and/or hardware platforms, mainly during the deployment activity. Please note, a *SepmDIPattern* does not have *TechnicalInterfaces*.
- *Property*. Is a particular characteristic of a pattern related to the concern it is dealing with and dedicated to capture its intent in a certain way. For instance, security properties. Each property of a pattern will be validated at the time of the pattern validating process and the assumptions used, will be compiled as a set of constraints which will have to be satisfied by the domain application.
- *SepmDSPattern*. Is a refinement of a *SepmDIPattern*. It is used to build a pattern at at domain specific level (DSPM). Furthermore a *SepmDSPattern* has *Technical Interfaces* in order to interact with the platform. This is the key entry artifact to model pattern at DSPM.

3.2 Specification Process

We propose an iterative specification process consisting of the following phases: (1) the specification of the pattern at domain independent level (DIPM), (2) the refinement of

DIPM pattern to specify one of its representations at domain specific level (DSPM). These two levels of the secure communication pattern presented in Section 2.1 are illustrated. For the sake of clarity, many functions and artifacts of this pattern have been omitted. We only detail the properties and interfaces that we need to describe the validation process. Note that the artifacts representing the formalization and the proof are detailed in Section 4.

The first step in our specification process is the understanding of the pattern's informal representation. The target representation is the SEPM metamodel, simplified in Figure 1 for the purpose of our study (semi-formal and formal representation of security pattern). The informal description given in Section 2.1 reflects our understanding from the representation of secure communication pattern given in literature [22]. At the DIPM level this description reveals the following elements: interfaces of type *SepmExternalInterface* and security properties of type *SepmProperties*. At the DSPM level, the description reveals the following elements: interfaces of type *SepmExternalInterface* and *SepmTechnicalInterface* and security properties of type *SepmProperties*. The description with varying levels of abstraction is managed by inheritance. Once there is a good understanding of the pattern informal representation structure, the pattern can be specified using the SEPM metamodel. The first step is to create an instance of *SepmPattern*. The instance is given a set of attributes representing the pattern. In our example, an instance of *SepmPattern* is created and named 'SecureCommPattern'.

Once the basic pattern subsystem has been specified, interfaces are added to expose some of the pattern's functionalities. For each such interface, an instance of *SepmExternalInterface* is added to the pattern's interfaces collection. The next step after creating interfaces is the creation of properties instances. An instance is created in the pattern's properties collection to specify every identified security property. A property is given a name and an expression on external interfaces in a property language. A complete specification of this pattern at both DIPM and DSPM levels is described below.

3.2.1 Domain Independent Pattern Model (DIPM).
At DIPM level, the security pattern subsystem and its related elements are created by inheritance. In our example, the DIPM of the secure communication pattern consists of:

- *External Interfaces.* The secure communication pattern exposes its functionalities through an external interface offering the following function calls:
  - $Send(P, Q, ch(P, Q), m)$,
  - $Receive(P, Q, ch(P, Q), m)$, with $P, Q \in \{C, S\}$, $ch(C, S) = ch(S, C)$ denoting the communication channel of client and server, and m a message.
- *Properties.* At this level, we specify the security property: "authenticity of sender and receiver". We denote this property as $auth(Send(C, S, ch(C, S), m)$, $Receive(S, C, ch(C, S), m), S)$ where the server S invokes the primitives $Send(C, S, ch(C, S), m)$ and $Receive(S, C, ch(C, S), m)$.

3.2.2 Domain Specific Pattern Model (DSPM).
At DSPM level, the security pattern and some of its related elements are also created by inheritance. Once a SEPMDSPATTERN is created, every pattern external interface is identified and modeled as a refinement of the SEPMEXTERNALINTERFACE in the pattern's interfaces collection. Then,

each of the pattern's technical interface is identified and modeled by an instance of *SepmTechnicalInterface* in the pattern's interfaces collection. The next step is the specification of properties. Each property is represented by an instance of *SepmProperty* in the pattern's properties collection. A property is given a name and an expression on external and technical interfaces in a property language.

For instance, when using SSL as a mechanism related to the application domain to refine the secure communication pattern at DSPM, we manage the following artifacts:

- *External Interfaces*. The DS external interface, a refinement of the DI external interface, can be specified as follows:

  - $send(C, S, mac_C(m), m)$: The client C sends m and the corresponding MAC (Message Authentication Code) to the server S.

  - $recv(S, C, mac_C(m), m)$: The server receives m and corresponding MAC.

- *Technical Interfaces*. The most important functions of the DS technical interface of the SSL pattern can be specified as follows:

  - $genRand(C, R_C), genRand(S, R_S)$: Client/server generate a random number.

  - $verifyCert()$: Client/server verify each others certificate and extract the respective public key.

  - $encrypt(C, pubKey_S, R_C)$: The client encrypts its random number using the server's public key.

  - $sign(C, ...)$: The client signs the SSL handshake messages.

  - $verifySig(S, ...)$: The server verifies the client's signature.

  - $genMac(C, macKey_C, m, mac_C(m))$: The client generates the MAC for message m using its own SSL shared secret for MAC generation.

  - $verifyMac(S, macKey_C, m, mac_C(m))$: The server verifies, using its shared secret for MAC verification (i.e. the client's key for MAC generation), that the MAC for m is correct and thus originates from the client.

  - $send(), recv()$: Send and receive of the SSL messages by client and server, respectively.

- *Properties*. In addition to the refinement of the authenticity property identified in the DIPM, at this level we may identify some related resource properties, e.g. the size of the cryptographic key.

4 Pattern Validation Process

We propose to use the interactive Isabelle/HOL proof assistant [17] for formalizing security patterns. The formal specification and verification of a pattern is represented through an Isabelle theory. It consists of a set of types, functions and theorems. In some situations, intermediate lemmas have been introduced in order to simplify what we have to prove.

We have defined a set of transformation rules to map the pattern metamodel concepts onto those of Isabelle/HoL. In our case, we focus on interfaces and properties. These interfaces provide functionalities through function calls that are encoded as actions whose parameters are agents and data. On the other side, pattern properties denote some expected outcomes over these actions. A property is given a name and an expression on agents and actions. Sometimes a property may refer to another property.

Each concept involved in the modeling of interfaces and properties is translated in Isabelle as an abbreviation, i.e. a new name for an existing construction, using data-types, records and definitions. For instance, the record *action* and the abbreviation *createAction* allow to create the *Send* and *Receive* actions. Following this definitional approach, we build the pattern specification from the bottom.

The designer can then extend this theory by verifying properties using the Isabelle scripting language for writing proofs using tactics. The goal now is to identify the set of assumptions required to prove the properties. Then, the proof consists to find a scheduling of valid steps using Isabelle's tactics such as applying a simplification considering that each step corresponds to a sub-goal to resolve (*command apply*). Our proof only uses simplification rules which consist in rewriting specified equations from left to right in the current goal (*command simp*).

Correctness of the proof is guaranteed by construction from the first goal to resolve (a lemma or a theorem) until the message "no sub-goals" is produced by the framework which confirms that the proof is finished (*command done*). For instance, to prove the authenticity property expressed as the goal *lemma authSendReceive* : *"auth Send Receive server"* with respect to the definition *auth*, each subgoal *auth*, *Send*, *Receive*, ... is rewritten according to its definition.

In the following, we apply the formalization and the verification processes to both DIPM and DSPM levels, and then we present the poof that the DIPM is an abstraction of the DSPM. The definitions introduced below are extracted from the code of our experiment. Definitions mainly concern the refinement process of SCP from the DIPM model to the DSPM model, using the SSL mechanism for the application domain.

4.1 Pattern Formalization

We first introduce the main concepts of authenticity, precedence and trust that are relevant for SCP. We call a particular action a authentic for an agent P after a sequence of actions if in all sequences that P considers to have possibly happened a must have happened. This is modeled by the $auth(a, b, P)$ predicate which denotes that whenever a particular action b has happened, it must be authentic for agent P that action a has happened as well. In the same way, $precede(a, b)$ holds if all action sequences in the system's behavior that contain an action b also contain an action a. Finally, $trust(P, prop)$ means that P trusts a property $prop$ to hold in the system if the property holds in the agent's conception of the system. The following gives types for these predicates in Isabelle.

$$auth ::" action => action => agent => bool"$$
$$precede ::" action => action => bool"$$
$$trust ::" agent => property => bool"$$

Before encoding the authenticity relation in Isabelle/HOL, one has to define what are an action, an agent and a property. All three are records i.e. tuples, with a name attached to each component. They will be independently introduced when defining the DIPM and DSPM pattern models respectively in Figure 2 and Figure 3.

```
 1: record action = agent1  :: "agentType"
                    agent2  :: "agentType"
                    channel :: "functorType"
                    message :: "messageType"
 2: definition createAction  :: "agentType==>agentType==>functorType==>messageType==>
action" where "createAction a1 a2 ca me==agent1=a1, agent2=a2, channel=ca, message=me"
 3: definition Send :: "action" where "Send == createAction C S ch m"
 4: definition Receive :: "action" where "Receive == createAction S C ch m"
 5: record property = action1 :: "action"
                      action2 :: "action"
 6: definition createProperty :: "action==>action==>property" where
"createProperty a1 a2 == action1=a1, action2=a2"
 7: definition precede :: "action==>action==>property" where
"precede a1 a2 == createProperty a1 a2"
 8: definition precedeSendReceive :: "property" where
"precedeSendReceive == precede Send Receive"
 9: record agent = name       :: "agentType"
                   actions     :: "action set"
                   properties  :: "property set"
10: definition createAgent :: "agentType==>(action set)==>(property
set)==>agent" where "createAgent a as ps == name=a, actions=as, properties=ps"
11: definition client :: "agent" where "client == createAgent C {Send} {}"
12: definition server :: "agent" where
"server == createAgent S {Send, Receive} {precedeSendReceive}"
13: definition auth :: "action==>action==>agent==>bool" where
"auth a1 a2 a == a1 IN (actions a) AND a2 IN (actions a)"
14: definition trust :: "agent==>property==>bool" where
"trust a p == p IN (properties a)"
15: definition authIfTrust :: "action==>action==>agent==>bool" where
"authIfTrust a1 a2 a == trust a (precede a1 a2)->auth a1 a2 a"
```

Fig. 2. The DIPM formalization of the Secure Communication Pattern

4.1.1 Domain Independent Pattern Model (DIPM). As previously defined in Section 3.2.1, the secure communication pattern exposes its functionalities through the two primitives *Send* and *Receive*. Applying the previous definition *auth*, we have to define these two primitives at the DIPM level. Figure 2 highlights the corresponding formalization in Isabelle owing to the *createAction* definition and the *Send* and *Receive* constants (lines 2-4). In the same way, an agent is introduced by the *createAgent* definition and the subject constant nominates the active computational entity of the pattern (line 10). The modeled cooperating system is so composed of a set of agents (e.g. some clients and a server) and a set of actions (e.g. the *Send* and *Receive* interface actions introduced above). We consider that the set of all possible sequences of actions models the behavior of the system and that an agent's initial knowledge about the system consists of all traces the agent initially considers possible. An agent may assume for example that a message that was received must have been sent before, and this is the case for any validation of a security property holding in a system. In our example, the SCP's DIPM level states that a *Receive* action from a client C using a specific channel ch is always preceded by the respective *Send* generation action triggered by the server S (line 8).

4.1.2 Domain Specific Pattern Model (DSPM). The formal model corresponding to the DSPM introduced in Section 3.2.2, as the refinement of the DIPM using the SSL mechanism, contains the same set of agents, namely C and S. Figure 3 depicts

```
1: record actionMac = agent   :: "agentType"
                       functor1 :: "functorType"
                       data    :: "messageType"
                       functor2 :: "functorType"
2: definition createActionMac :: "agentType==>functorType==>messageType ==>
functorType ==> actionMac" where "createActionMac a f1 d f2 == (agent=a,
functor1=f1, data=d, functor2=f2)"
3: definition genMac :: "actionMac" where
"genMac == createActionMac C macKey m mac"
4: definition verifyMac :: "actionMac" where "verifyMac == createActionMac S
macKey m mac"
5: record propertyMac = action1 :: "actionMac"
                        action2 :: "actionMac"
6: definition createPropertyMac :: "actionMac==>actionMac==>propertyMac" where
"createPropertyMac a1 a2 == (action1=a1, action2=a2)"
7: definition precede :: "actionMac ==>actionMac==>propertyMac" where
"precede a1 a2 == createPropertyMac a1 a2"
8: definition precedeGenMacVerifyMac :: "propertyMac" where
"precedeGenMacVerifyMac == precede genMac verifyMac"
9: record agent =   name :: "agentType"
                    actionsMac :: "actionMac set"
                    actionsRandom :: "actionRandom set"
                    actionsKey :: "actionKey set"
                    propertiesMac :: "propertyMac set"
                    propertiesRandom :: "propertyRandom set"
                    propertiesKey :: "propertyKey set"
10: definition createAgent :: "agentType==>(actionMac set)==>(actionRandom set)
==>(actionKey set)==>(propertyMac set)==>(propertyRandom set)==>(propertyKey
set)==>agent" where "createAgent a asm asr ask psm psr psk == (name=a, actionsMac=asm,
actionsRandom=asr, actionsKey=ask, propertiesMac=psm, propertiesRandom=psr,
propertiesKey=psk)"
11: definition server :: "agent" where "server == createAgent S {genMac,verifyMac}
{genRand} {privKeyS, privKeyCA} {precedeGenMacVerifyMac} {notPrecedeGenRandGenRand}
{confPrivKeyS, confPrivKeyCA}"
12: definition trustMac :: "agent==>propertyMac==>   bool" where
"trust a p ==  p IN (propertiesMac a)"
```

Fig. 3. The DSPM formalization of the Secure Communication Pattern

some of these DSPM artifacts using the Isabelle definitions. Its actions correspond to the external and technical DS interface function calls. They can be considered as a refinement of the actions of the DIPM formal model. The code defines also a specific DSPM server S suitable for the SSL protocol (line 11). The security property that is provided by this SSL pattern and that corresponds to the trust property assumed to hold for the DIPM model is that the server trusts into the precedence of its own MAC verification action by the MAC generation action of the client (line 12).

4.2   Pattern Validation

Now we present the verification of the secure communication pattern at both DIPM and DSPM levels.

4.2.1   Domain Independent Pattern Model (DIPM). The DIPM formal model of Figure 2 defines the SCP's collaborative agents of the pattern refined at Figure 3 using the SSL mechanism. The former figure also introduces the *Send* and *Receive* actions corresponding to the external DI interface (lines 3 and 4). We further assume that each agent can only see its own actions. According to Section 3.2.1, the required authenticity property is expressed as: *lemma authSendReceive* : *"auth Send Receive server"* (P-DI)

In order to prove this constraint between *Send* and *Receive* messages, we use the result of [5] which states that the properties $trust(P, precede(a, b))$ and $auth(b, c, P)$ imply the property $auth(a, c, P)$. Setting $b = c$ and instantiating a, b, P with the concrete *Send* and *Receive* actions and the server S of property P-DI, we conclude that P-DI holds if the following properties (assumptions) hold:

- *lemma trustSendReceive* : *"trust server precedeSendReceive"* (A-DI)
- *lemma authReceiveReceive* : *"auth Receive Receive server"*

We may assume that the latter property holds because any action identified by the server is authentic. Hence, this concludes our proof with respect to the DIPM model the assumption (A-DI) must be assumed to hold.

We emulate these constraints in Isabelle by the theorem:

theorem authIfTrustSendReceive :   *"authIfTrust Send Receive server"*

which terminates the P-DI proof of the DIPM model assuming the assumption A-DI. For simplification purpose, we consider that trust of the server into the precedence of a corresponding *Receive* action by a client *Send* action is given by the set membership relation of the same action to the set of actions of mapped agents. Recall that Figure 2 introduces the definitions which encodes the precedence, auth and trust properties between a *Send* action and a *Receive* action for the couple client and server (lines 8, 13, 14 and 15).

The fact that no more DIPM definition can be applied shows that we now have to consider the DSPM level, i.e. we have to find and validate the SSL implementation that provides an equivalent property. This will be discussed in the next paragraph.

4.2.2   Domain Specific Pattern Model (DSPM). The security property that is provided by this SSL pattern and that corresponds to the trust property assumed to hold for the DIPM model is that the server trusts into the precedence of its own signature verification action by the signature generation action of the client:

lemma trust : *"trust server precedeGenMacVerifyMac"* (P-DS)

In order to prove P-DS, we isolate assumptions provided by the SSL protocol as a random number is only generated once, the server trusts into the confidentiality of its own private RSA key and into the confidentiality of the certificate authority's private key. These statements captures the semantics of an RSA encryption and allows to conclude that S trusts in the confidentiality of the shared secrets derived from the SSL handshake and yields that indeed property P-DS holds. For more details, the complete proof was introduced in [11]. In our case, we introduce the *genMac* and *verifyMac* actions and their temporal dependency *precedeGenMacVerifyMac*, as illustrated by Figure 3 (lines 3, 4 and 8). We also defined the trust predicate for the DSPM server (line 12). This predicate will be used by the refinement process from the DIPM model to the DSPM model (line 1 of Figure 4).

We synthesize the P-DS proof by introducing the *genMac* and *verifyMac* actions and their temporal dependency *precedeGenMacVerifyMac*, as illustrated by Figure 3 (lines 3, 4 and 8). However we need what is trust for a DSPM server; this is defined by the code of line 12. This predicate will be used by the refinement process from the DIPM model to the DSPM model (see line 1 of Figure 4).

4.3 Correspondence between DIPM and DSPM

Note that a system specification does not require a particular level of abstraction. Different formal models of the same system are partially ordered with respect to different levels of abstraction. Formally, abstractions can be mapped to action sequences of a finer abstraction level to action sequences of a more abstract level while respecting concatenation of actions.

Correspondence between the DIPM and DSPM formal models is assumed when proving that the property introduced at the DIPM model is transferred to the DPSM model. More precisely, this means that the DIPM model is an abstraction of the DSPM model. In particular, we must show that using a specific mechanism for verifying the property together with function calls of the specific domain is a specific case of proving the upper-level property.

We so have to map actions of the DSPM model onto the actions of the DIPM model by an appropriate homomorphism h and then prove that this homomorphism preserves *trust* in *precede*. In practice, h is required to preserve each operation or a pseudo-operation which summarizes the behavior of a set of operations. Figure 4 specifies h in Isabelle and the resulting $trustWithH$ theorem.

```
1:definition h :: "(DSPM.action * DIPM.action) list" where
"h == [(genMac, Send), (verifyMac, Receive)]"
2:definition buildProperty :: "(DSPM.action * DIPM.action) list==>DSPM.property"
where "buildProperty l == DSPM.property.action1=fst(nth l 0),
DSPM.property.action2=fst(nth l 1)"
3:definition applyH :: "(DSPM.action * DIPM.action) list==>DIPM.property" where
"applyH l ==DIPM.property.action1=snd (nth l 0), DIPM.property.action2=snd(nth l 1)"
4:definition buildAndApplyH :: "DSPM.agent==>DIPM.agent==>
(DSPM.action * DIPM.action) list==>bool" where
"buildAndApplyH a1 a2 l == DSPM.trust a1(buildProperty l)->
DIPM.trust a2 (applyH l)"
5:theorem trustWithH : "buildAndApplyH DSPM.server DIPM.server h"
```

Fig. 4. Proving *trust* in *precede* for the Secure Communication Pattern

Since we assume property A-DI to hold for the DIPM model, all server *Receive* actions are preceded by a client *Send* action in the server's abstract initial knowledge. On the other hand, the server's concrete initial knowledge reflects the MAC mechanism, i.e. reflects that a *verifyMac* action is always preceded by the respective *genMac* action. This is assumed by the *buildProperty* definition of Figure 4 (line 2) introduced as assumption and by the *precedeGenMacVerifyMac* definition of Figure 3 (line 8).

Note that we consider P-DS as the premise of an implication where $DSPM.trust$ $a1$ *(buildProperty l)* refers to *trust* for the DSPM model of Figure 3, while $DIPM.trust$ $a2$ *(applyH l)* refers to *trust* for the DIPM counterpart of Figure 2. More generally, the DIPM and DSPM call or type prefixes of Figure 4 are related to each modeling. Thereby, setting up and proving the $trustWithH$ theorem (line 5) requires defining both a DIPM server and a DSPM server; these two servers have the same functionality in practice. Based on these assumptions, the homomorphism h preserves trust into precedence and property A-DI transferred to the DSPM model is identical to property P-DS.

5   Related Works

Design patterns are a solution model to generic design problems, applicable in specific contexts. Supporting research tackles the presented challenges includes domain patterns, pattern languages and recently formalisms and modeling languages to foster their application in practice. To give an idea of the improvement achievable by using specific languages for the specification of patterns, we look at pattern formalization and modeling problems targeting the integration of the pattern specification and validation steps into a broader MDE process.

Several tentatives exist in the literature to deal with patterns for specific concerns [8,23]. They allow to solve very general problems that appear frequently as sub-tasks in the design of systems with security and dependability requirements. These elementary tasks include secure communication, fault tolerance, etc. The pattern specification consists of a service-based architectural design and deployment restrictions in form of UML deployment diagrams for the different architectural services.

To give an overview of the improvement achievable by using specific languages, we look at the pattern specification and formalization problems. UMLAUT [9] is an approach that aims to formally model design patterns by proposing extensions to the UML metamodel 1.3. They used OCL language to describe constraints (structural and behavioral) in the form of meta-collaboration diagrams. In the same way, RBML (Role-Based Metamodeling Language) [15] is able to capture various design perspectives of patterns such as static structure, interactions, and state-based behavior.

While many patterns for specific concern have been designed, still few works propose general techniques for patterns. For the first kind of approaches [6], design patterns are usually represented by diagrams with notations such as UML objects, annotated with textual descriptions and examples of code. There are some well-proven approaches [3] based on Gamma et al. However, this kind of technique does not allow to reach the high degree of pattern structure flexibility which is required to reach our target.

Formal specification has also been introduced in [16] in order to give rigorous reasoning of behavioral features of a design pattern in terms of high-level abstractions of communication. In this paper, the author considers an object-oriented formalism for reactive system (DisCo) [13] based on TLA (Temporal Logic of Actions) to express high-level abstractions of cooperation between objects involved in a design pattern. However, patterns are directly formalized at the pattern level including its classes, its relations and its actions, without defining a modeling language.

The work in [2] introduces a new specification template inspired on secure system development needs. The template is augmented with UML notations for the solution and with formal artifacts for the requirement properties. Another approach [7] provides a formal and visual language for specifying design patterns called LePUS. It defines a pattern in an accurate and complete form of formula in Z, with a graphical representation. The framework promoted by LePUS is interesting but the degree of expressiveness proposed to design a pattern is too restrictive.

UMLsec [14] is an approach based on the integration of modeling security in UML and formal methods for object-oriented system development. UMLsec is defined in form of a UML profile and semantics to assist in the automated analysis of the UMLsec models with respect to security requirements. UMLsec and our approach are not in competition but they complement each other by providing different view points to the secure information system, mainly in applying security patterns for system security engineering.

Moreover, [12] used the concept of security problem frames as analysis patterns for security problems and associated solution approaches. The analysis activities using these patterns are described with a highlight of how the solution may be set, with a focus on the privacy requirement anonymity. For software architecture, [10] presented an evaluation of security patterns in the context of secure software architectures. The evaluation is based on the existing methods for secure software development, such as guidelines as well as on threat categories.

To summarize, in software engineering, design patterns are considered effective tools for the reuse of specific knowledge. However, a gap between the development of systems using patterns and the pattern information still exists. This becomes even more visible when dealing with specific concerns namely security and dependability properties for several application sectors such as presented recently in [4]. In an other point of view, we agree with the argumentations given in [24] to justify why the precise specification and formalization of a pattern by definition restricts its "degree of freedom for the design", and hence there is no success stories of works dealing with pattern formalization. This is not only related to security patterns. Note however, that this work does not address the validation activity which is an important issue in any design activity and more particularly in security engineering. We think that security is subject to rigorous and precise specification and the proposed literature (in our best knowledge) fails to meet these two objectives. To remedy these contradictory needs, we support the specifications of security patterns at two levels of abstractions, domain independent and domain specific, in both a semi-formal representation through metamodeling techniques and a rigorous formal representation through interactive theorem proving approach. This allows to support some variability of the pattern.

6 Conclusion and Future Work

A classical form of pattern is not sufficient to tame the complexity of safety critical systems – complexity occurs because of both the concerns and the domain management. To reach this objective and to foster reuse, we introduced the specification at domain independent and domain specific levels. The former exhibits an abstract solution without specific knowledge on how the solution is implemented with regard to the application domain. Following an MDE process, the domain independent model of patterns is then refined towards a domain specific level, taking into account domain artifacts, concrete elements such as mechanisms to use, devices that are available, etc. Consequently, a security pattern at domain specific level contains the respective information.

These two levels of abstractions are captured using new concepts related to the different kind of knowledge described by the pattern had; not with existing software constructs. In our work, we used the MDE philosophy. We do not use the software concepts (object or component constructs) recommended by Model-Driven Architecture (MDA), for example. However, the SEPM language is subject to target specific software modeling languages such as those recommended by MDA, using model transformation techniques. Regarding the well known MDA levels (PIM and PSM), there is an overlap between these levels and our two abstraction levels. For example, in the Intelligent Transport Systems (ITS) domain, the ISO/IEC 15118 highly recommends to use TLS for ensuring security properties. For that domain, we can find different platforms supporting such a DSPM pattern.

We also provide an accompanying formalization and validation framework to help precise specification of patterns based on the interactive Isabelle/HOL proof assistant. The resulting validation artifact's may mainly (1) complete the definitions, and (2) provide semantics for the interfaces and the properties in the context of S&D. Like this, validation artifacts may be added to the pattern for traceability concerns. In the same way, the domain refinement is applied during the formal validation process for the specification and validation of patterns.

Furthermore, we walk through a prototype of EMF tree-based editors supporting the approach. Currently the tool suite named *Semcomdt*[2] is provided as Eclipse plugins. The approach presented here has been evaluated on two case studies from the TERESA project[3] resulting in the development of a repository of S&D patterns with more than 30 S&D patterns.

The next step of this work consists in defining a correct-by-construction pattern-based security engineering process. It aims to provide the correct-by-construction integration of a design pattern into an application while offering a certain degree of liberty to the designer using it. In order to be able to validate the integration, we must have a formal specification of the pattern, i.e., its properties, constraints and related validation artifacts, as input to the pattern-based development process. Another objective for the near future is to provide automated tool support for pattern-based development, preferably based on a widely known and accepted model-based approach in industry such as UML [18]. For that, we plan to investigate the possibility to transform these design artifacts into UML [18] and their corresponding validation artifacts into OCL [19].

References

1. Alexander, C., Ishikawa, S., Silverstein, M.: A Pattern Language. Center for Environmental Structure Series, vol. 2. Oxford University Press, New York (1977)
2. Cheng, B., Cheng, B.H.C., Konrad, S., Campbell, L.A., Wassermann, R.: Using security patterns to model and analyze security. In: IEEE Workshop on Requirements for High Assurance Systems, pp. 13–22 (2003)
3. Douglass, B.P.: Real-time UML: Developing Efficient Objects for Embedded Systems. Addison-Wesley (1998)

---

[2] http://www.semcomdt.org
[3] http://www.teresa-project.org/

4. Fernandez, E.B., Yoshioka, N., Washizaki, H., Jürjens, J., VanHilst, M., Pernul, G.: Software Engineering for Secure Systems: Industrial and Research Perspectives. In: Mouratidis, H. (ed.) IGI Global, pp. 16–31 (2010)

5. Fuchs, A., Gürgens, S., Rudolph, C.: A Formal Notion of Trust – Enabling Reasoning about Security Properties. In: Nishigaki, M., Jøsang, A., Murayama, Y., Marsh, S. (eds.) IFIPTM 2010. IFIP AICT, vol. 321, pp. 200–215. Springer, Heidelberg (2010)

6. Gamma, E., Helm, R., Johnson, R.E., Vlissides, J.: Design Patterns: Elements of Reusable Object-Oriented Software. Addison-Wesley (1995)

7. Gasparis, E., Nicholson, J., Eden, A.H.: LePUS3: An Object-Oriented Design Description Language. In: Stapleton, G., Howse, J., Lee, J. (eds.) Diagrams 2008. LNCS (LNAI), vol. 5223, pp. 364–367. Springer, Heidelberg (2008)

8. Di Giacomo, V., et al.: Using Security and Dependability Patterns for Reaction Processes. In: International Workshop on Database and Expert Systems Applications, pp. 315–319. IEEE Computer Society (2008)

9. Le Guennec, A., Sunyé, G., Jézéquel, J.-M.: Precise modeling of design patterns. In: Evans, A., Caskurlu, B., Selic, B. (eds.) UML 2000. LNCS, vol. 1939, pp. 482–496. Springer, Heidelberg (2000)

10. Halkidis, S.T., Chatzigeorgiou, A., Stephanides, G.: A qualitative analysis of software security patterns. Computers & Security 25(5), 379–392 (2006)

11. Hamid, B., Gürgens, S., Jouvray, C., Desnos, N.: Enforcing S&D Pattern Design in RCES with Modeling and Formal Approaches. In: Whittle, J., Clark, T., Kühne, T. (eds.) MODELS 2011. LNCS, vol. 6981, pp. 319–333. Springer, Heidelberg (2011)

12. Hatebur, D., Heisel, M., Schmidt, H.: A security engineering process based on patterns. In: Proceedings of the 18th International Conference on Database and Expert Systems Applications, DEXA 2007, pp. 734–738. IEEE Computer Society, Washington, DC (2007)

13. Jarvinen, H.M., Kurki-Suonio, R.: DisCo specification language: marriage of actions and objects. In: 11th International Conference on Distributed Computing Systems, pp. 142–151. IEEE Press (1991)

14. Jürjens, J.: UMLsec: Extending UML for Secure Systems Development. In: Jézéquel, J.-M., Hussmann, H., Cook, S. (eds.) UML 2002. LNCS, vol. 2460, pp. 412–425. Springer, Heidelberg (2002)

15. Kim, D.K., France, R., Ghosh, S., Song, E.: A UML-Based Metamodeling Language to Specify Design Patterns. In: Patterns, Proc. Workshop Software Model Eng (WiSME) with Unified Modeling Language Conf. 2004, pp. 1–9 (2004)

16. Mikkonen, T.E.: Formalizing design patterns. In: Proceeding ICSE 1998 Proceedings of the 20th International Conference on Software Engineering. IEEE Press (1998)

17. Nipkow, T., Paulson, L.C., Wenzel, M.T.: Isabelle/HOL. LNCS, vol. 2283. Springer, Heidelberg (2002)

18. OMG. OMG Unified Modeling Language (OMG UML), Superstructure (February 2009), http://www.omg.org/spec/UML/2.2/Superstructure

19. OMG. OCL 2.2 Specification (February 2010)

20. Schmidt, D.: Model-Driven Engineering. IEEE Computer 39(2), 41–47 (2006)

21. Schumacher, M.: Security Engineering with Patterns. LNCS, vol. 2754. Springer, Heidelberg (2003)

22. Schumacher, M., Fernandez-Buglioni, E., Hybertson, D., Buschmann, F., Sommerlad, P.: Security Patterns: Integrating Security and Systems Engineering. Wiley Software Patterns Series. John Wiley & Sons (2006)

23. Yoshioka, N., Washizaki, H., Maruyama, K.: A survey of security patterns. Progress in Informatics (5), 35–47 (2008)

24. Zdun, U., Avgeriou, P.: Modeling Architectural Patterns Using Architectural Primitives. In: Proceedings of the 20th Annual ACM SIGPLAN Conference on Object-oriented Programming, Systems, Languages, and Applications, OOPSLA 2005, pp. 133–146. ACM, New York (2005)

25. Zhang, T., Jouault, F., Bezivin, J., Zhao, J.: A MDE Based Approach for Bridging Formal Models. In: Sixth International Symposium on Theoretical Aspects of Software Engineering, pp. 113–116 (2008)

26. Zurawski, R.: Embedded Systems. In: Embedded Systems Handbook. CRC Press Inc. (2005)

A. Isabelle/HOL Formalization and Validation of the Secure Communication Pattern

The complete formalisation and verification of the secure communication pattern, using Isabelle/HoL, are available online via http://www.semcomdt.org/semco/resources/SCP_Isabelle.zip.

A.1. DIPM Validation

```
theory DIPM imports Main
begin
(* Secure Communication DIPM Pattern definitions *)
...
definition authIfTrust :: "action ==> action ==> agent ==> bool" where
"authIfTrust a1 a2 a == trust a (precede a1 a2) —> auth a1 a2 a"

(* Secure Communication DIPM Pattern validation *)
theorem authIfTrustSendReceive : "authIfTrust Send Receive server"
apply (simp only : authIfTrust_def)
apply (simp only : trust_def)
apply (simp only : server_def)
apply (simp only : createProperty_def createAgent_def)
apply (simp only : Send_def Receive_def)
apply (simp only : createAction_def)
apply (simp only : auth_def)
apply (simp)
done
end
```

A.2. DSPM Formalization and Validation

```
theory DSPM imports Main
begin
(* Secure Communication DSPM Pattern definitions *)
...
definition trustRandom :: "agent==>propertyRandom ==>bool" where
"trustRandom a p == p IN (propertiesRandom a)"
definition trustKey :: "agent ==> propertyKey ==> bool" where
"trustKey a p == p IN (propertiesKey a)"
definition trustInConfidentiality :: "agent==> bool" where
"trustInConfidentiality a == (trustRandom a notPrecedeGenRandGenRand) AND
(trustKey a confPrivKeyS) AND (trustKey a confPrivKeyCA)"
definition trustInSignature :: "agent==> bool" where
"trustInSignature a == (trustMac a (precede genMac verifyMac))"
definition trustSCP :: "agent ==> bool" where
"trustSCP a == trustInConfidentiality a —> trustInSignature a"

(* Secure Communication DSPM Pattern validation *)
theorem proofTrustSCP : "trustSCP server"
apply (simp only : trustSCP_def)
apply (simp only : trustInConfidentiality_def trustInSignature_def)
apply (simp only : notPrecedeGenRandGenRand_def confPrivKeyS_def confPrivKeyCA_def)
```

```
apply (simp only : genMac_def verifyMac_def)
apply (simp only : notPrecede_def precede_def conf_def)
apply (simp only : createPropertyMac_def createActionMac_def)
apply (simp only : createPropertyRandom_def createActionRandom_def)
apply (simp only : createPropertyKey_def createActionKey_def)
apply (simp only : trustMac_def trustRandom_def trustKey_def)
apply (simp only : genRand_def privKey1_def privKey2_def)
apply (simp only : server_def)
apply (simp only : createActionMac_def    createActionRandom_def
createActionKey_def createAgent_def)
apply (simp)
apply (simp only : notPrecedeGenRandGenRand_def confPrivKeyS_def confPrivKeyCA_def)
apply (simp only : precedeGenMacVerifyMac_def)
apply (simp only : precede_def notPrecede_def conf_def)
apply (simp only : createPropertyMac_def   createPropertyRandom_def createPropertyKey_def)
apply (simp only : genMac_def verifyMac_def privKeyS_def privKeyCA_def genRand_def)
apply (simp only : createActionMac_def createActionRandom_def createActionKey_def)
apply (simp)
done
end
```

A.3. Correspondence between DIPM and DSPM Formalization and Validation

```
theory DIPMtoDSPM imports DIPM DSPM
begin
(* From DIPSM Pattern to DSPM Pattern definitions *)
...
definition buildAndApplyH :: "DSPM.agent ==> DIPM.agent ==>
(DSPM.action * DIPM.action) list ==> bool" where
"buildAndApplyH a1 a2 l == DSPM.trust a1 (buildProperty l) ->
DIPM.trust a2 (applyH l)"


(* From DIPSM Pattern to DSPM Pattern validation *)
theorem trustWithH : "buildAndApplyH DSPM.server DIPM.server h"
apply (simp only : buildAndApplyH_def)
apply (simp only : DIPM.trust_def DSPM.trust_def)
apply (simp only : buildProperty_def)
apply (simp only : applyH_def)
apply (simp)
apply (simp only : h_def)
apply (simp only : Send_def Receive_def)
apply (simp only : DIPM.createAction_def)
apply (simp only : genMac_def verifyMac_def)
apply (simp only : DSPM.createAction_def)
apply (simp only : DIPM.server_def DSPM.server_def)
apply (simp)
apply (simp only : DIPM.createAgent_def DSPM.createAgent_def)
apply (simp only : Send_def Receive_def genMac_def verifyMac_def)
apply (simp)
apply (simp only : precedeSendReceive_def precedeGenMacVerifyMac_def)
apply (simp only : DIPM.precede_def DIPM.createProperty_def DSPM.precede_def
DSPM.createProperty_def)
apply (simp)
apply (simp only : Send_def Receive_def genMac_def verifyMac_def)
apply (simp only : DIPM.createAction_def DSPM.createAction_def)
apply (simp)
done
end
```

On the Relation between Redactable and Sanitizable Signature Schemes

Hermann de Meer[1,3], Henrich C. Pöhls[2,3,*],
Joachim Posegga[2,3], and Kai Samelin[4,**]

[1] Chair of Computer Networks and Computer Communication
[2] Chair of IT-Security
[3] Institute of IT-Security and Security Law (ISL), University of Passau, Germany
[4] Engineering Cryptographic Protocols Group & CASED, TU Darmstadt, Germany
demeer@uni-passau.de, {hp,jp}@sec.uni-passau.de, kai.samelin@ec-spride.de

Abstract. Malleable signature schemes ($\mathcal{MSS}$) enable a third party to alter signed data in a controlled way, maintaining a valid signature after an authorized change. Most well studied cryptographic constructions are (1) redactable signatures ($\mathcal{RSS}$), and (2) sanitizable signatures ($\mathcal{SSS}$). $\mathcal{RSS}$s allow the removal of blocks from a signed document, while $\mathcal{SSS}$s allow changing blocks to arbitrary strings. We rigorously prove that $\mathcal{RSS}$s are less expressive than $\mathcal{SSS}$s: no unforgeable $\mathcal{RSS}$ can be transformed into an $\mathcal{SSS}$. For the opposite direction we give a black-box transformation of a single $\mathcal{SSS}$, with tightened security, into an $\mathcal{RSS}$.

1 Introduction

Digital signatures are *the* IT-Security mechanism applied to detect integrity violations, as they become invalid on any change to the signed data. However, this also prohibits third parties from changing signed data in an allowed and controlled way. Applications where such an alteration is crucial, include secure routing [2] or "blank signatures" [23]. An additional prevailing reason to allow for subsequent changing or removing parts is the anonymization of personally identifiable information (PII), e.g., in medical data [28,39]. Apart from the important privacy guarantee for the original data, it is often of paramount importance that the action of modification requires no additional interaction with the original signer. Hence, they are applicable in a wide area, e.g., in the Internet-of-Things (IoT) or for cloud computing [30]. For example, consider the IoT: communication

---

* The research leading to these results has received funding from the European Union's Seventh Framework Programme (FP7/2007-2013) under grant agreement n° 609094.

** Work partly carried out while working at the University of Passau, supported by "Regionale Wettbewerbsfähigkeit und Beschäftigung", Bayern, 2007-2013 (EFRE) as part of the SECBIT project (www.secbit.de) and the European Community's Seventh Framework Programme through the EINS Network of Excellence under grant agreement n° 288021. This work was also partly supported by the German Federal Ministry of Education and Research (BMBF) within EC SPRIDE and by the Hessian LOEWE excellence initiative within CASED.

J. Jürjens, F. Piessens, and N. Bielova (Eds.): ESSoS 2014, LNCS 8364, pp. 113–130, 2014.
© Springer International Publishing Switzerland 2014

with the originating sensor for re-signing would dramatically increase communication costs. Also in the case of smart metering privacy [16], the originating Smart meter must not know in what way the data it signed is later modified to preserve the user's privacy.

One cryptographically suitable approach to solve the above described "digital document sanitization problem" [34] are malleable signature schemes. Malleable signature schemes authorize certain changes to signed data such that the resulting changed message's authenticity is still verifiable, i.e., it remains verifiable that either none or only authorized changes have been applied to the signed message. This authorized change can come in several facets: Let $m = (m[1], \ldots, m[\ell])$, where $\ell \in \mathbb{N}$ and $m[i] \in \{0,1\}^*$, be a string m split up into ℓ parts we refer to as *blocks*. First, redactable signature schemes $(\mathcal{RSS})$ allow *anyone* to remove blocks $m[i]$ from m, without invalidating the signature. In particular, a redaction of the block $m[i]$, $0 < i \leq \ell$ leaves a blinded message m' without $m[i]$, i.e., $m' = (\ldots, m[i-1], \square, m[i+1], \ldots)$. A *visible* $\square$ has a major impact on the $\mathcal{RSS}$'s privacy guarantees. For $\mathcal{RSS}$s, it is required that *anyone* can derive a signature σ' which verifies for m'. Second, sanitizable signature schemes $(\mathcal{SSS})$, allow a sanitizer to change the *admissible* blocks, which are predefined by the signer, into arbitrary strings $m[i]' \in \{0,1\}^*$. Hence, the sanitizer can generate a verifiable message-signature pair (m', σ'). Contrary to $\mathcal{RSS}$s, in $\mathcal{SSS}$s the sanitizer holds its own secret key. Obviously, it must be verifiable that all changes were endorsed by the signer in both concepts.

Motivation. When more and more data containers are digitally signed as a countermeasure against attacks on their integrity and authenticity, it becomes increasingly important to be able to remove contained sensitive data with little impact on the integrity of the remaining data. In other words, signing data becomes the standard solution to allow integrity checks, e.g., for data stored on third-party storage servers in the cloud [30]. Current provably secure (unforgeable and private) solutions mostly focus on one out of two specific types of malleable signature schemes, i.e., either on $\mathcal{RSS}$s *or* $\mathcal{SSS}$s, often stating the other as related work, or even the same. At first sight, both approaches aim for the same goal, i.e., changing signed data. In detail, $\mathcal{SSS}$s only allow for *alterations* of blocks, while $\mathcal{RSS}$s only allow *removal of complete* blocks. Moreover, $\mathcal{SSS}$s require an additional key pair, while $\mathcal{RSS}$s have public redactions. The questions we answer in this paper is: What is the exact relation between both types of malleable signature?

Findings. We prove that an $\mathcal{RSS}$ is not *trivially* a "special case" of $\mathcal{SSS}$ [8,41,42]. But first things first: Obviously an unforgeable $\mathcal{SSS}$ can emulate a standard signature by disallowing any modifications by any sanitizer [42]. Second, we note that a $\mathcal{RSS}$ for a message of n blocks, fulfilling privacy [7], can trivially be constructed by deploying $\mathcal{O}(n^2)$ standard digital signatures [7,37]. Hence, $\mathcal{O}(n^2)$ $\mathcal{SSS}$s are sufficient to construct one $\mathcal{RSS}$. Thus, from a theoretical point of view, $\mathcal{SSS}$s directly imply the existence of $\mathcal{RSS}$s. However, from a practical point of view, such constructions are rather inefficient. Especially as $\mathcal{RSS}$ can be constructed in $\mathcal{O}(n)$ for computation and storage [37]. We provide the security definitions required to transform only a *single* invocation of an $\mathcal{SSS}$ into an $\mathcal{RSS}$. We prove that the existing security models are not sufficient to achieve such a transformation. In particular, the resulting $\mathcal{RSS}$s cannot fulfill the state-of-the-art privacy definitions, as introduced in [7]. Note, for all transforms we treat $\mathcal{SSS}$ and $\mathcal{RSS}$ as *black-boxes* and ignore the constructions' details.

Contribution. This paper rigorously shows that $\mathcal{RSS}$s do not imply $\mathcal{SSS}$s: no unforgeable $\mathcal{RSS}$ can be transformed into a secure $\mathcal{SSS}$. While the converse is true in general, we give a more detailed separation: one *cannot* construct a *fully secure* $\mathcal{RSS}$ from a single $\mathcal{SSS}$ invocation, if one treats the used $\mathcal{SSS}$ as a black-box. In detail, weakening the privacy definition of $\mathcal{RSS}$s, while strengthening the security definitions of $\mathcal{SSS}$s, a single invocation of such a strong $\mathcal{SSS}$ can emulate a weaker $\mathcal{RSS}$. This paper provides an algorithm for a general transform: Any $\mathcal{SSS}$ that is (1) strongly private, (2) weakly immutable and (3) weakly blockwise non-interactive publicly accountable can be transformed into a weakly private $\mathcal{RSS}$. We give formal definitions of all the security properties in Sect. 2. Interestingly enough, it turns out that our definition of weak privacy is fulfilled by many existing $\mathcal{RSS}$s, which are considered not private following the model by *Brzuska* et al. [7].

Related Work. $\mathcal{SSS}$s have been introduced by *Ateniese* et al. [2] at ESORICS '05. *Brzuska* et al. formalized the most essential security properties [8]. These have later been extended for the properties of unlinkability [10,12] and (block/groupwise) non-interactive public accountability [11,18]. Moreover, several extensions and modifications like limiting-to-values [13,27,36], trapdoor $\mathcal{SSS}$s [15] and multi-sanitizer environments [14] have been considered.

$\mathcal{RSS}$s were introduced in 2002 by *Johnson* et al. in [26]. In the same year, *Steinfeld* and *Bull* introduced a similar concept as "Content Extraction Signatures" [40]. Since then, $\mathcal{RSS}$s have been subject to much research and got extended to tree-structured data [7,28] and to arbitrary graphs [29]. *Samelin* et al. introduced the concept of redactable structure in [38]. The standard security properties of $\mathcal{RSS}$s have been formalized in [7,17,37]. *Ahn* et al. introduced the notion of context-hiding $\mathcal{RSS}$s [1]. Even stronger privacy notions have recently been introduced in [3,4]. However, the scheme by *Ahn* et al. only achieves the less common notion of *selective* unforgeability [1]. Moreover, [1,3,4] are limited to quoting, i.e., redactions are only possible at the beginning, or end resp., of a list. There exists many additional work on $\mathcal{RSS}$s. We do note that most of the schemes are not fully private, e.g., [22,24,25,31,33]. Hence, a verifier can make statements about the original message m, which contradicts the intention of an $\mathcal{RSS}$ [7]. Most of these schemes achieve our notion of "weak privacy".

Combinations of both approaches appeared in [22,24,25]. However, their schemes do not preserve privacy [38]. While the work of *Yum* and *Joong* tries to combine the two properties in [42], the authors are not aware of any work considering relations between the notions.

Malleable signature schemes are usable in practice according to [35,36].

We do note that there are also schemes aiming for calculating general functions on signed data, e.g., [5,6,19]. In this work, we focus on the relation between $\mathcal{SSS}$s and $\mathcal{RSS}$s.

2 Preliminaries and Security of $\mathcal{SSS}$ and $\mathcal{RSS}$

For a message $m = (m[1], \ldots, m[\ell])$, we call $m[i] \in \{0,1\}^*$ a *block*, where "," denotes a uniquely reversible concatenation of blocks or strings. The symbol $\perp \notin \{0,1\}^*$ denotes an error or an exception. For a visible redaction, we use the symbol $\square \notin \{0,1\}^*$, $\square \neq \perp$.

Sanitizable Signatures. The used notation is adapted from [8].

Definition 1 (Sanitizable Signature Scheme). *A $\mathcal{SSS}$ consists of at least seven efficient (PPT) algorithms $\mathcal{SSS} := (\mathsf{KGen}_{sig}, \mathsf{KGen}_{san}, \mathsf{Sign}, \mathsf{Sanit}, \mathsf{Verify}, \mathsf{Proof}, \mathsf{Judge})$:*

Key Generation. *There are two key generation algorithms, one for the signer and one for the sanitizer. Both create a pair of keys, a private key and the public key, using the security parameter λ:*
$$(\mathrm{pk}_{sig}, \mathrm{sk}_{sig}) \leftarrow \mathsf{KGen}_{sig}(1^\lambda), \quad (\mathrm{pk}_{san}, \mathrm{sk}_{san}) \leftarrow \mathsf{KGen}_{san}(1^\lambda)$$

Signing. *The Sign algorithm takes $m = (m[1], \ldots, m[\ell])$, $m[i] \in \{0,1\}^*$, the signer's secret key sk_{sig}, the sanitizer's public key pk_{san}, as well as a description ADM of the admissibly modifiable blocks, where ADM contains the number ℓ of blocks in m, as well the indices of the modifiable blocks. It outputs the message m and a signature σ (or $\perp$, indicating an error):*
$$(m, \sigma) \leftarrow \mathsf{Sign}(1^\lambda, m, \mathrm{sk}_{sig}, \mathrm{pk}_{san}, \mathrm{ADM})$$

Sanitizing. *Algorithm Sanit takes a message $m = (m[1], \ldots, m[\ell])$, $m[i] \in \{0,1\}^*$, a signature σ, the public key pk_{sig} of the signer and the secret key sk_{san} of the sanitizer. It modifies the message m according to the modification instruction MOD, which contains pairs $(i, m[i]')$ for those blocks that shall be modified. Sanit calculates a new signature σ' for the modified message $m' \leftarrow \mathrm{MOD}(m)$. Then Sanit outputs m' and σ' (or $\perp$, indicating an error):*
$$(m', \sigma') \leftarrow \mathsf{Sanit}(1^\lambda, m, \mathrm{MOD}, \sigma, \mathrm{pk}_{sig}, \mathrm{sk}_{san})$$

Verification. *The Verify algorithm outputs a decision $d \in \{\mathit{true}, \mathit{false}\}$ verifying the validity of a signature σ for a message $m = (m[1], \ldots, m[\ell])$, $m[i] \in \{0,1\}^*$ with respect to the public keys:*
$$d \leftarrow \mathsf{Verify}(1^\lambda, m, \sigma, \mathrm{pk}_{sig}, \mathrm{pk}_{san})$$

Proof. *The Proof algorithm takes as input the security parameter, the secret signing key sk_{sig}, a message $m = (m[1], \ldots, m[\ell])$, $m[i] \in \{0,1\}^*$ and a signature σ as well a set of (polynomially many) additional message-signature*

pairs $\{(m_i, \sigma_i) \mid i \in \mathbb{N}\}$ and the public key pk_{san}. It outputs a string $\pi \in \{0,1\}^*$ (or $\perp$, indicating an error):
$$\pi \leftarrow \text{Proof}(1^\lambda, \text{sk}_{sig}, m, \sigma, \{(m_i, \sigma_i) \mid i \in \mathbb{N}\}, \text{pk}_{san})$$

Judge. Algorithm **Judge** takes as input the security parameter, a message $m = (m[1], \ldots, m[\ell])$, $m[i] \in \{0,1\}^*$ and a valid signature σ, the public keys of the parties and a proof π. It outputs a decision $d \in \{Sig, San, \perp\}$ indicating whether the message-signature pair has been created by the signer or the sanitizer (or $\perp$, indicating an error): $d \leftarrow \text{Judge}(1^\lambda, m, \sigma, \text{pk}_{sig}, \text{pk}_{san}, \pi)$

To have an algorithm actually able to derive the accountable party for a specific block $m[i]$, Brzuska et al. introduced the additional algorithm Detect [11]. The algorithm Detect is not part of the original SSS description by Ateniese et al., since it is not required for the purpose of a SSS [2,8]. However, we require this algorithm later on to define (weak) blockwise non-interactive public accountability (See Def. 6).

Definition 2 (SSS Detect). On input of the security parameter λ, a message-signature pair (m, σ), the corresponding public keys pk_{sig} and pk_{san}, and a block index $1 \leq i \leq \ell$, Detect outputs the accountable party (**San** or **Sig**) for block i (or $\perp$, indicating an error):
$$d \leftarrow \text{Detect}(1^\lambda, m, \sigma, \text{pk}_{sig}, \text{pk}_{san}, i), d \in \{San, Sig, \perp\}$$

We require the usual correctness properties to hold. In particular, all genuinely signed or sanitized messages are accepted, while every genuinely created proof π by the signer leads the judge to decide in favor of the signer. For a formal definition of correctness, refer to [8,11]. It is also required by every SSS that ADM is always correctly recoverable from any valid message-signature pair (m, σ). This accounts for the work done in [21]. Jumping ahead, we want to emphasize that an SSS with weak non-interactive public accountability requires that Judge detects any *sanitization* on input of an empty proof $\pi = \perp$. Formal definitions of the security properties in a game-based manner follow.

Redactable Signatures. The following notation is derived from [38].

Definition 3 (Redactable Signature Schemes). An RSS consists of four efficient algorithms $RSS := (\text{KeyGen}, \text{Sign}, \text{Verify}, \text{Redact})$:

KeyGen. The algorithm **KeyGen** outputs the public key pk and private key sk of the signer, where λ denotes the security parameter:
$$(\text{pk}, \text{sk}) \leftarrow \text{KeyGen}(1^\lambda)$$

Sign. The algorithm **Sign** gets as input the secret key sk and the message $m = (m[1], \ldots, m[\ell])$, $m[i] \in \{0,1\}^*$: $(m, \sigma) \leftarrow \text{Sign}(1^\lambda, \text{sk}, m)$

Verify. The algorithm **Verify** outputs a decision $d \in \{true, false\}$, indicating the validity of the signature σ, w.r.t. pk, protecting $m = (m[1], \ldots, m[\ell])$, $m[i] \in \{0,1\}^*$: $d \leftarrow \text{Verify}(1^\lambda, \text{pk}, m, \sigma)$

Experiment Unforgeability$_{\mathcal{RSS},\mathcal{A}}(\lambda)$
  $(pk, sk) \leftarrow \mathsf{KeyGen}(1^\lambda)$
  $(m^*, \sigma^*) \leftarrow \mathcal{A}^{\mathsf{Sign}(1^\lambda, sk, \cdot)}(pk)$
    let $i = 1, \ldots, q$ denote the queries to Sign
    return 1, if
      $\mathsf{Verify}(1^\lambda, pk, m^*, \sigma^*) = 1$ and
      for all $i = 1, \ldots, q : m^* \notin \mathrm{span}_{\vDash}(m_i)$

Fig. 1. Unforgeability for $\mathcal{RSS}$

Redact. *The algorithm* Redact *takes as input the message* $m = (m[1], \ldots, m[\ell])$, $m[i] \in \{0, 1\}^*$, *the public key* pk *of the signer, a valid signature* σ *and a list of indices* MOD *of blocks to be redacted. It returns a modified message* $m' \leftarrow \mathrm{MOD}(m)$ *(or* $\bot$, *indicating an error):*
$$(m', \sigma') \leftarrow \mathsf{Redact}(1^\lambda, \mathrm{pk}, m, \sigma, \mathrm{MOD})$$
We denote the transitive closure of m *as* $\mathrm{span}_{\vDash}(m)$. *This set contains all messages derivable from* m *w.r.t.* Redact

As for $\mathcal{SSS}$s, the correctness properties for $\mathcal{RSS}$s are required to hold as well. Thus, every genuinely signed or redacted message must verify. Refer to [7] for a formal definition of correctness.

Security Models. This section contains the required security properties and models. They are derived from [8,21,38], but have been significantly altered. The requirement that ADM is always correctly reconstructible is captured within the unforgeability and immutability definitions. Note, following [8,11,12], an $\mathcal{SSS}$ must at least be unforgeable, immutable, accountable and private to be meaningful. Hence, we assume that all used $\mathcal{SSS}$s fulfill these four fundamental security requirements; if these requirements are not met, the construction is not considered an $\mathcal{SSS}$ and the results of this paper are not directly applicable. On the other hand, an $\mathcal{RSS}$ must be unforgeable and (weakly) private to be meaningful [7].

Unforgeability. No one should be able to compute a valid signature on a message not previously issued without having access to any private keys [7]. This is analogous to the unforgeability requirement for standard signature schemes [20], except that it excludes valid redactions from the set of forgeries for $\mathcal{RSS}$s, while for $\mathcal{SSS}$s *no* alterations are allowed.

Definition 4 ($\mathcal{RSS}$ Unforgeability). *We say that an* $\mathcal{RSS}$ *is unforgeable, if for any efficient (PPT) adversary* $\mathcal{A}$ *the probability that the game depicted in Fig. 1 returns 1, is negligible (as a function of* λ*).*

Definition 5 ($\mathcal{SSS}$ Unforgeability). *We say an* $\mathcal{SSS}$ *is unforgeable, if for any efficient (PPT) adversary* $\mathcal{A}$ *the probability that the game depicted in Fig. 2 returns 1, is negligible (as a function of* λ*).*

Experiment Unforgeability$_{SSS,\mathcal{A}}(\lambda)$

$(pk_{\text{sig}}, sk_{\text{sig}}) \leftarrow \text{KGen}_{\text{sig}}(1^\lambda)$

$(pk_{\text{san}}, sk_{\text{san}}) \leftarrow \text{KGen}_{\text{san}}(1^\lambda)$

$(m^*, \sigma^*) \leftarrow \mathcal{A}^{\text{Sign}(1^\lambda, \cdot, sk_{\text{sig}}, \cdots)\text{Proof}(1^\lambda, \cdot, sk_{\text{sig}}, \cdots), \text{Sanit}(1^\lambda, \cdots, sk_{\text{san}})}(pk_{\text{sig}}, pk_{\text{san}})$

   let $(m_i, \text{ADM}_i, pk_{\text{san},i})$ and σ_i for $i = 1, 2, \ldots q$

   denote the queries/answers to/by the oracle Sign,

   let $(m_j, \text{MOD}_j, \sigma_j, pk_{\text{sig},j})$ and (m'_j, σ'_j) for $j = q+1, \ldots, r$

   denote the queries/answers to/by the oracle Sanit.

   return 1, if

   Verify$(1^\lambda, m^*, \sigma^*, pk_{\text{sig}}, pk_{\text{san}}) = \text{true}$ and

   for all $q = 1, \ldots q : (pk_{\text{san}}, m^*, \text{ADM}^*) \neq (pk_{\text{san},i}, m_i, \text{ADM}_i)$ and

   for all $j = q+1, \ldots, r : (pk_{\text{sig}}, m^*, \text{ADM}^*) \neq (pk_{\text{sig},j}, m_i, \text{ADM}_i)$

Fig. 2. Unforgeability for SSS

Experiment WBlockPubAcc$_{SSS,\mathcal{A}}(\lambda)$

$(pk_{\text{sig}}, sk_{\text{sig}}) \leftarrow \text{KGen}_{\text{sig}}(1^\lambda)$

$(pk_{\text{san}}, sk_{\text{san}}) \leftarrow \text{KGen}_{\text{san}}(1^\lambda)$

$(m^*, \sigma^*) \leftarrow \mathcal{A}^{\text{Sign}(1^\lambda, \cdot, sk_{\text{sig}}, pk_{\text{san}}, \cdot), \text{Proof}(1^\lambda, \cdot, sk_{\text{sig}}, \cdots, pk_{\text{san}})}(pk_{\text{san}}, sk_{\text{san}}, pk_{\text{sig}})$

   let (m_i, ADM_i) and (m_i, σ_i) for $i = 1, \ldots, k$ be queries/answers to/by Sign

   return 1, if

   Verify$(1^\lambda, m^*, \sigma^*, pk_{\text{sig}}, pk_{\text{san}}) = \text{true}$ and

   $\exists q$, s.t. Detect$(1^\lambda, m^*, \sigma^*, pk_{\text{sig}}, pk_{\text{san}}, q) = \text{Sig}$ and

   for all $i = 1, \ldots, k : (m^*[q], \sigma^*) \neq (m_i[q], \sigma_i)$.

return 0

Fig. 3. Weak Blockwise Non-Interactive Public Accountability for SSS

Weak Blockwise Non-Interactive Public Accountability. The basic idea is that an adversary, i.e., the sanitizer, has to be able to make the Detect algorithm accuse the signer, if it did not sign the specific block. Moreover, in our definition, the signer is *not* considered adversarial, contrary to *Brzuska* et al. [11]. An example for a weakly blockwise non-interactive publicly accountable SSS is the scheme introduced by *Brzuska* et al. [11]. We explain the reasons for our adversary model after the introduction of all required security properties. Note, pk_{san} is fixed for the oracles. For SSSs, we also have sanitization and proof oracles [8].

Definition 6 (SSS Weak Blockwise Non-interactive Public Account-ability). *A sanitizable signature scheme SSS is* weakly non-interactive publicly accountable, *if Proof $= \perp$, and if for any efficient algorithm $\mathcal{A}$ the probability that the experiment given in Fig. 3 returns 1 is negligible (as a function of λ).*

Privacy. No one should be able to gain any knowledge about sanitized parts without having access to them [8]. This is similar to the standard indistinguishability notion for encryption schemes. The basic idea is that the oracle either signs and sanitizes the first message (m_0) or the second (m_1), while the resulting message must be the same for each input. The adversary must not be able to decide which input message was used.

Experiment $\text{Privacy}_{SSS,\mathcal{A}}(\lambda)$

$\quad (pk_{\text{sig}}, sk_{\text{sig}}) \leftarrow \text{KGen}_{\text{sig}}(1^\lambda)$

$\quad (pk_{\text{san}}, sk_{\text{san}}) \leftarrow \text{KGen}_{\text{san}}(1^\lambda)$

$\quad b \leftarrow \{0,1\}$

$\quad a \leftarrow \mathcal{A}^{\text{Sign}(1^\lambda, sk_{\text{sig}}, \cdots), \text{Proof}(1^\lambda, sk_{\text{sig}}, \cdots), \text{LoRSanit}(\cdots, sk_{\text{sig}}, sk_{\text{san}}, b), \text{Sanit}(1^\lambda, \cdots, sk_{\text{san}})}(pk_{\text{sig}}, pk_{\text{san}})$

$\quad\quad$ where oracle LoRSanit on input of:

$\quad\quad m_0, \text{MOD}_0, m_1, \text{MOD}_1, \text{ADM}$

$\quad\quad$ if $\text{MOD}_0(m_0) \neq \text{MOD}_1(m_1)$, return $\perp$

$\quad\quad$ if $\text{MOD}_0 \not\subseteq \text{ADM} \vee \text{MOD}_1 \not\subseteq \text{ADM}$, return $\perp$

$\quad\quad$ let $(m, \sigma) \leftarrow \text{Sign}(1^\lambda, m_b, sk_{\text{sig}}, pk_{\text{san}}, \text{ADM})$

$\quad\quad$ return $(m', \sigma') \leftarrow \text{Sanit}(1^\lambda, m, \text{MOD}_b, \sigma, pk_{\text{sig}}, sk_{\text{san}})$

$\quad$ return 1, if $a = b$

Fig. 4. Standard Privacy for SSS

Definition 7 (SSS Standard Privacy). *We say that an SSS is (standard) private, if for any efficient (PPT) adversary $\mathcal{A}$ the probability that the game depicted in Fig. 4 returns 1, is negligibly close to $\frac{1}{2}$ (as a function of λ).*

The aforementioned privacy definition [8] only considers outsiders as adversarial. However, we require that even insiders, i.e., sanitizers, are not able to win the game. Note, the key sk_{san} is *not* generated by the adversary, only known to it. We explain the need for this alteration after the next definitions. For our definition of strong privacy, the basic idea remains the same: no one should be able to gain any knowledge about sanitized parts without having access to them, with one exception: the adversary is given the secret key sk_{san} of the sanitizer. This notion extends the definition of standard privacy (Fig. 4) to also account for parties knowing the secret sanitizer key sk_{san}. In a sense, this definition captures some form of "forward-security". Examples for strongly private SSSs are the schemes introduced by *Brzuska* et al. [9,11,12], as their schemes are perfectly private. As the adversary now knows sk_{san}, it can trivially simulate the sanitization oracle itself.

Definition 8 (SSS Strong Privacy). *We say that an SSS is private, if for any efficient (PPT) adversary $\mathcal{A}$ the probability that the game depicted in Fig. 5 returns 1, is negligibly close to $\frac{1}{2}$ (as a function of λ).*

In a weakly private RSS, a third party can derive which parts of a message have been redacted without gathering more information, as redacted blocks are replaced with $\square$, which is visible. The basic idea is that the oracle either signs and sanitizes the first message (m_0) or the second (m_1). As before, the resulting redacted message m' must be the same for both inputs, with one additional exception: the length of both inputs must be the same, while $\square$ is considered part of the message. For strong privacy, this constraint is not required. We want to emphasize, that *Lim* et al. define weak privacy in a different manner: they prohibit access to the signing oracle [31]. Our definition allows for such adaptive queries. Summarized, weak privacy only makes statements about blocks, not the complete message. See [28] for possible attacks. Weakly private schemes, following our definition, are, e.g., [22,28]. In their schemes, the adversary is able to pinpoint the indices of the redacted blocks, as $\square$ is visible.

Experiment $\mathsf{SPrivacy}_{SSS,\mathcal{A}}(\lambda)$
$\quad (pk_{\mathrm{sig}}, sk_{\mathrm{sig}}) \leftarrow \mathsf{KGen}_{\mathrm{sig}}(1^\lambda)$
$\quad (pk_{\mathrm{san}}, sk_{\mathrm{san}}) \leftarrow \mathsf{KGen}_{\mathrm{san}}(1^\lambda)$
$\quad b \leftarrow \{0, 1\}$
$\quad a \leftarrow \mathcal{A}^{\mathsf{Sign}(1^\lambda, \cdot, sk_{\mathrm{sig}}, pk_{\mathrm{san}}, \cdot), \mathsf{Proof}(1^\lambda, sk_{\mathrm{sig}}, \cdots, pk_{\mathrm{san}}), \mathsf{LoRSanit}(\cdots, sk_{\mathrm{sig}}, sk_{\mathrm{san}}, b)}(pk_{\mathrm{sig}}, pk_{\mathrm{san}}, sk_{\mathrm{san}})$
$\quad\quad$ where oracle $\mathsf{LoRSanit}$ on input of:
$\quad\quad m_0, \mathrm{MOD}_0, m_1, \mathrm{MOD}_1, \mathrm{ADM}$
$\quad\quad$ if $\mathrm{MOD}_0(m_0) \neq \mathrm{MOD}_1(m_1)$, return $\bot$
$\quad\quad$ if $\mathrm{MOD}_0 \not\subseteq \mathrm{ADM} \vee \mathrm{MOD}_1 \not\subseteq \mathrm{ADM}$, return $\bot$
$\quad\quad$ let $(m, \sigma) \leftarrow \mathsf{Sign}(1^\lambda, m_b, sk_{\mathrm{sig}}, pk_{\mathrm{san}}, \mathrm{ADM})$
$\quad\quad$ return $(m', \sigma') \leftarrow \mathsf{Sanit}(1^\lambda, m, \mathrm{MOD}_b, \sigma, pk_{\mathrm{sig}}, sk_{\mathrm{san}})$
$\quad$ return 1, if $a = b$

Fig. 5. Strong Privacy for SSS

Experiment $\mathsf{WPrivacy}_{RSS,\mathcal{A}}(\lambda)$
$\quad (pk, sk) \leftarrow \mathsf{KeyGen}(1^\lambda)$
$\quad b \leftarrow \{0, 1\}$
$\quad d \leftarrow \mathcal{A}^{\mathsf{Sign}(1^\lambda, sk, \cdot), \mathsf{LoRRedact}(\cdots, sk, b)}(pk)$
$\quad\quad$ where oracle $\mathsf{LoRRedact}$
$\quad\quad$ for input $m_0, m_1, \mathrm{MOD}_0, \mathrm{MOD}_1$:
$\quad\quad$ if $\mathrm{MOD}_0(m_0) \neq \mathrm{MOD}_1(m_1)$, return $\bot$
$\quad\quad$ Note: redacted blocks are denoted $\square$, which are considered part of m
$\quad\quad (m, \sigma) \leftarrow \mathsf{Sign}(1^\lambda, sk, m_b)$
$\quad\quad$ return $(m', \sigma') \leftarrow \mathsf{Redact}(1^\lambda, pk, m, \sigma, \mathrm{MOD}_b)$.
$\quad$ return 1, if $b = d$

Fig. 6. Weak Privacy for RSS

Definition 9 (RSS Weak Privacy). *We say that an RSS is weakly private, if for any efficient (PPT) adversary $\mathcal{A}$ the probability that the game depicted in Fig. 6 returns 1, is negligibly close to $\frac{1}{2}$ (as a function of λ).*

The next definition is similar to weak privacy. However, redacted parts are *not* considered part of the message.

Definition 10 (RSS Strong Privacy). *We say that an RSS is strongly private, if for any efficient (PPT) adversary $\mathcal{A}$ the probability that the game depicted in Fig. 7 returns 1, is negligibly close to $\frac{1}{2}$ (as a function of λ). This is the standard definition of privacy [7].*

Immutability. The idea behind immutability is that an adversary generating the sanitizer key must only be able to sanitize admissible blocks. Hence, immutability is the unforgeability requirement for the sanitizer.

Definition 11 (SSS Immutability). *A sanitizable signature scheme SSS is immutable, if for any efficient algorithm $\mathcal{A}$ the probability that the experiment from Fig. 8 returns 1 is negligible (as a function of λ) [8].*

For weak immutability, an adversary knowing, but not generating, the sanitizer key must only be able to sanitize admissible blocks. Hence, once more, pk_{san} is fixed.

Experiment SPrivacy$_{\mathcal{RSS},\mathcal{A}}(\lambda)$
$(pk, sk) \leftarrow \mathsf{KeyGen}(1^\lambda)$
$b \leftarrow \{0, 1\}$
$d \leftarrow \mathcal{A}^{\mathsf{Sign}(1^\lambda, sk, \cdot), \mathsf{LoRRedact}(\cdots, sk, b)}(pk)$
    where oracle LoRRedact
    for input $m_0, m_1, \mathrm{MOD}_0, \mathrm{MOD}_1$:
    if $\mathrm{MOD}_0(m_0) \neq \mathrm{MOD}_1(m_1)$, return $\bot$
    Note: redacted blocks are *not* considered part of the message
    $(m, \sigma) \leftarrow \mathsf{Sign}(1^\lambda, sk, m_b)$
    return $(m', \sigma') \leftarrow \mathsf{Redact}(1^\lambda, pk, m, \sigma, \mathrm{MOD}_b)$.
return 1, if $b = d$

Fig. 7. Strong Privacy for $\mathcal{RSS}$

Experiment Immutability$_{\mathcal{SSS},\mathcal{A}}(\lambda)$
$(pk_{\mathrm{sig}}, sk_{\mathrm{sig}}) \leftarrow \mathsf{KeyGen}(1^\lambda)$
$(m^*, \sigma^*, pk^*) \leftarrow \mathcal{A}^{\mathsf{Sign}(1^\lambda, \cdot, sk_{\mathrm{sig}}, \cdot, \cdot), \mathsf{Proof}(1^\lambda, sk_{\mathrm{sig}}, \cdots)}(pk_{\mathrm{sig}})$
let $(m_i, \mathrm{ADM}_i, pk_{\mathrm{san},i})$ and σ_i for $i = 1, \ldots, q$ be queries/answers to/by Sign
return 1, if:
    $\mathsf{Verify}(1^\lambda, m^*, \sigma^*, pk_{\mathrm{sig}}, pk^*) = \mathbf{true}$ and
    for all $i = 1, 2, \ldots, q : (pk^*, m^*[j_i], \mathrm{ADM}^*) \neq (pk_{\mathrm{san},i}, m_i[j_i], \mathrm{ADM}_i)$ and
    if $(m^*[j_i], \mathrm{ADM}, pk_{\mathrm{san},i}) \neq (m_i[j_i], \mathrm{ADM}_i, pk_{\mathrm{san},i})$, also $j_i \notin \mathrm{ADM}_i$
    where shorter messages are padded with $\bot$

Fig. 8. Immutability for $\mathcal{SSS}$

Definition 12 ($\mathcal{SSS}$ Weak Immutability). *A sanitizable signature scheme $\mathcal{SSS}$ is weakly immutable, if for any efficient algorithm $\mathcal{A}$ the probability that the experiment given in Fig. 9 returns 1 is negligible (as a function of λ).*

Interestingly, weak immutability is enough for our construction to be unforgeable, while for an $\mathcal{RSS}$ used in the normal way, this definition is obviously not suitable at all due to accountability reasons. We omit the security parameter λ for the rest of the paper to increase readability.

Implications and Separations. Let us formulate our first theorems:

Theorem 1. *There exists an $\mathcal{RSS}$ which is only weakly private.*

Proof. See [22,24,25,31] for examples.

Theorem 2. *Every $\mathcal{SSS}$ which is immutable, is also weakly immutable.*

Proof. Trivially implied: $\mathcal{A}$ generates the sanitizer key pair honestly.

Theorem 3. *There exists an $\mathcal{SSS}$ which is private, but not strongly private.*

Th. 3 is proven in App. B.

Experiment $\mathsf{WImmutability}_{SSS,\mathcal{A}}(\lambda)$
  $(pk_{\text{sig}}, sk_{\text{sig}}) \leftarrow \mathsf{KeyGen}(1^\lambda)$
  $(pk_{\text{san}}, sk_{\text{san}}) \leftarrow \mathsf{KeyGen}(1^\lambda)$
  $(m^*, \sigma^*) \leftarrow \mathcal{A}^{\mathsf{Sign}(1^\lambda, \cdot, sk_{\text{sig}}, pk_{\text{san}}, \cdot), \mathsf{Proof}(1^\lambda, sk_{\text{sig}}, \cdots, pk_{\text{san}}, \cdot)}(pk_{\text{sig}}, pk_{\text{san}}, sk_{\text{san}})$
  let (m_i, ADM_i) and σ_i for $i = 1, 2, \ldots q$ be queries/answers to/by Sign
  return 1, if:
    $\mathsf{Verify}(1^\lambda, m^*, \sigma^*, pk_{\text{sig}}, pk_{\text{san}}) = \mathbf{true}$ and
    $\forall i, i = 1, 2, \ldots, q : (m^*[j_i], \text{ADM}^*) \neq (m_i[j_i], \text{ADM}_i)$ and
    if $(m^*[j_i], \text{ADM}^*) \neq (m_i[j_i], \text{ADM}_i)$, also $j_i \notin \text{ADM}_i$
    where shorter messages are padded with $\bot$

Fig. 9. Weak Immutability for SSS

Definition of a Secure $\mathcal{RSS}$ and a Secure SSS. We want to explicitly emphasize that accountability, as defined for SSSs in [8], has not been defined for $\mathcal{RSS}$s yet, as Redact is a public algorithm. Hence, no secret sanitizer key(s) are required for redactions. To circumvent this inconsistency, we utilize a standard SSS and let the signer generate the sanitizer key sk_{san}, attaching it to the public key of the signer. This also explains why pk_{san} is fixed in our security model. If any alteration without sk_{san} is possible, the underlying SSS would obviously be forgeable. As we have defined that this is a non-secure SSS, we omit this case. Hence, the secret sk_{san} becomes public knowledge and can be used by every party. This is the reason why the adversary only knows sk_{sig}, but cannot generate it. We require these, at first sight very unnatural, restrictions to stay consistent with the standard model of SSSs as formalized in [8]. Moreover, the signer is generally *not* considered an adversarial entity in $\mathcal{RSS}$s [7]. If other notions or adversary models are used, the results may obviously differ. In App. A, we show that any SSS which only achieves standard privacy, is not enough to construct a weakly private $\mathcal{RSS}$ and additional impossibility results.

3 Generic Transformation

This section presents the generic transform. In particular, we provide a generic algorithm which transforms any weakly immutable, strongly private, and weakly blockwise non-interactive publicly accountable SSS into an unforgeable and weakly private $\mathcal{RSS}$.

Outline. The basic idea of our transform is that every party, including the signer, is allowed to alter *all* given blocks. The verification procedure accepts sanitized blocks, if the altered blocks are $\square$. $\square$ is treated as a redacted block. Hence, redaction is altering a given block to a special symbol. As we have defined that an SSS only allows for strings $m[i] \in \{0, 1\}^*$, we need to define $\square := \emptyset$ and $m[i] \leftarrow 0$, if $m[i] = \emptyset$ and $m[i] \leftarrow m[i] + 1$ else to codify the additional symbol $\square$. Here, $\emptyset$ expresses the empty string. Hence, we remain in the model defined. Moreover, this is where weak blockwise non-public interactive public accountability comes in: the changes to *each* block need to be detectable to allow for a meaningful result, as an SSS allows for arbitrary alterations. As $\square$

is still visible, the resulting scheme is only weakly private, as statements about m can be made. This contradicts our definition of strong privacy for $\mathcal{RSS}$s. Moreover, as an $\mathcal{RSS}$ allows every party to redact blocks, it is obvious that sk_{san} must be known to every party, including the signer. Therefore, we need a strongly private $\mathcal{SSS}$ to achieve our definition of weak privacy for the $\mathcal{RSS}$, as proven in App. A.

Construction 1. *Let* $\mathcal{SSS} := (KGen_{sig}, KGen_{san}, Sign, Sanit, Verify, Proof, Judge, Detect)$ *be a secure* $\mathcal{SSS}$. *Define* $\mathcal{RSS} := (KeyGen, Sign, Verify, Redact)$ *as follows:*

Key Generation: Algorithm KeyGen *generates on input of the security param-*
eter λ, *a key pair* $(pk_{sig}, sk_{sig}) \leftarrow \mathcal{SSS}.KGen_{sig}(1^\lambda)$ *of the* $\mathcal{SSS}$, *and also a*
sanitizer key pair $(pk_{san}, sk_{san}) \leftarrow \mathcal{SSS}.KGen_{san}(1^\lambda)$. *It returns* $(sk, pk) =$
$(sk_{sig}, (sk_{san}, pk_{san}, pk_{sig}))$
Signing: Algorithm $\mathcal{RSS}$.Sign *on input* $m \in \{0,1\}^*, sk, pk,$ *sets* ADM $= (1, \ldots, \ell)$
and computes $\sigma \leftarrow \mathcal{SSS}.Sign(1^\lambda, m, sk_{sig}, pk_{san}, \text{ADM})$. *It outputs:* (m, σ)
Redacting: Algorithm $\mathcal{RSS}$.Redact *on input message* m, *modification instruc-*
tions MOD, *a signature* σ, *keys* pk $= (sk_{san}, pk_{san}, pk_{sig})$, *first checks if* σ
is a valid signature for m *under the given public keys using* $\mathcal{RSS}$.Verify.
If not, it stops outputting $\bot$. *Afterwards, it sets* MOD' $= \{(i, \Box) \mid i \in$
MOD$\}$. *In particular, it generates a modification description for the* $\mathcal{SSS}$
which sets block with index $i \in$ MOD *to* $\Box$. *Finally, it outputs* $(m', \sigma') \leftarrow$
$\mathcal{SSS}.Sanit(1^\lambda, m, \text{MOD}', \sigma, pk_{sig}, sk_{san})$
Verification: Algorithm $\mathcal{RSS}$.Verify *on input a message* $m \in \{0,1\}^*$, *a signature*
σ *and* pk *first checks that* ADM $= (1, \ldots, \ell)$ *and that* σ *is a valid signature*
for m *under the given public keys using* $\mathcal{SSS}$.Verify. *If not, it returns* **false**.
Afterwards, for each i *for which* $\mathcal{SSS}.Detect(1^\lambda, m, \sigma_s, pk_{sig}, pk_{san}, i)$ *returns*
San, *it checks that* $m[i] = \Box$. *If not, it returns* **false**. *Else, it returns* **true**.
One may also check, if sk_{san} *is correct and that all* $m[i]$ *are sanitizable, if*
required.

Theorem 4 (Our Construction is Secure). *If the utilized* $\mathcal{SSS}$ *is weakly blockwise non-interactive publicly accountable, weakly immutable and strongly private, the resulting* $\mathcal{RSS}$ *is weakly private, but not strongly, and unforgeable.*

Th. 4 is proven in App. B.

As $\mathcal{RSS}$s allow for removing every block, we require that ADM $= (1, \ldots, \ell)$. This rules out cases where a signer prohibits alterations of blocks. This constraint can easily be transformed into the useful notion of consecutive disclosure control [32,38].

4   Conclusion and Future Work

This paper presents a method to transform a single instantiation of an $\mathcal{SSS}$ into an $\mathcal{RSS}$. In detail, if we use one $\mathcal{SSS}$ instantiation, an emulation of an $\mathcal{RSS}$ can only be achieved, if the $\mathcal{SSS}$'s security is strengthened, raising it above the existing standard. The resulting emulated $\mathcal{RSS}$ offers only weaker privacy guarantees. Moreover, we have argued rigorously that the opposite implication is not possible. Thus, no $\mathcal{RSS}$ can be transformed into an unforgeable $\mathcal{SSS}$. Hence, $\mathcal{RSS}$s and $\mathcal{SSS}$s are indeed two different cryptographic building blocks, even if they achieve to define and delegate authorized modifications of signed messages. Currently, the number of $\mathcal{SSS}$s achieving the new security requirements needed to securely emulate an $\mathcal{RSS}$ is still low.

For the future, we suggest to focus on implementing and standardizing an $\mathcal{SSS}$ secure enough to emulate $\mathcal{RSS}$s, to have one universal building block. In the meantime we advice to use dedicated $\mathcal{RSS}$ algorithms if only redactions are needed and a $\mathcal{SSS}$ algortihm. Of course, you are advised to check current work to ensure the cryptographic strength of the constructions.

Cryptographically, remaining open questions are: how to formally define accountability for $\mathcal{RSS}$s, to identify if the interesting privacy properties of unlinkability for $\mathcal{SSS}$ [10,12] will carry forward when transformed into an $\mathcal{RSS}$, and to further research how $\mathcal{RSS}$ and $\mathcal{SSS}$s can be combined.

References

1. Ahn, J.H., Boneh, D., Camenisch, J., Hohenberger, S., Shelat, A., Waters, B.: Computing on authenticated data. Cryptology ePrint Archive, Report 2011/096 (2011), http://eprint.iacr.org
2. Ateniese, G., Chou, D.H., de Medeiros, B., Tsudik, G.: Sanitizable signatures. In: de Capitani di Vimercati, S., Syverson, P.F., Gollmann, D. (eds.) ESORICS 2005. LNCS, vol. 3679, pp. 159–177. Springer, Heidelberg (2005)
3. Attrapadung, N., Libert, B., Peters, T.: Computing on authenticated data: New privacy definitions and constructions. In: Wang, X., Sako, K. (eds.) ASIACRYPT 2012. LNCS, vol. 7658, pp. 367–385. Springer, Heidelberg (2012)
4. Attrapadung, N., Libert, B., Peters, T.: Efficient completely context-hiding quotable and linearly homomorphic signatures. In: Kurosawa, K., Hanaoka, G. (eds.) PKC 2013. LNCS, vol. 7778, pp. 386–404. Springer, Heidelberg (2013)
5. Boneh, D., Freeman, D.M.: Homomorphic signatures for polynomial functions. In: Paterson, K.G. (ed.) EUROCRYPT 2011. LNCS, vol. 6632, pp. 149–168. Springer, Heidelberg (2011)
6. Boneh, D., Freeman, D.M.: Linearly homomorphic signatures over binary fields and new tools for lattice-based signatures. In: Catalano, D., Fazio, N., Gennaro, R., Nicolosi, A. (eds.) PKC 2011. LNCS, vol. 6571, pp. 1–16. Springer, Heidelberg (2011)
7. Brzuska, C., et al.: Redactable Signatures for Tree-Structured Data: Definitions and Constructions. In: Zhou, J., Yung, M. (eds.) ACNS 2010. LNCS, vol. 6123, pp. 87–104. Springer, Heidelberg (2010)
8. Brzuska, C., Fischlin, M., Freudenreich, T., Lehmann, A., Page, M., Schelbert, J., Schröder, D., Volk, F.: Security of Sanitizable Signatures Revisited. In: Jarecki, S., Tsudik, G. (eds.) PKC 2009. LNCS, vol. 5443, pp. 317–336. Springer, Heidelberg (2009)
9. Brzuska, C., Fischlin, M., Lehmann, A., Schröder, D.: Sanitizable signatures: How to partially delegate control for authenticated data. In: Proc. of BIOSIG. LNI, vol. 155, pp. 117–128. GI (2009)
10. Brzuska, C., Fischlin, M., Lehmann, A., Schröder, D.: Unlinkability of Sanitizable Signatures. In: Nguyen, P.Q., Pointcheval, D. (eds.) PKC 2010. LNCS, vol. 6056, pp. 444–461. Springer, Heidelberg (2010)

11. Brzuska, C., Pöhls, H.C., Samelin, K.: Non-Interactive Public Accountability for Sanitizable Signatures. In: De Capitani di Vimercati, S., Mitchell, C. (eds.) EuroPKI 2012. LNCS, vol. 7868, pp. 178–193. Springer, Heidelberg (2013)
12. Brzuska, C., Pöhls, H.C., Samelin, K.: Efficient and Perfectly Unlinkable Sanitizable Signatures without Group Signatures. In: Agudo, I. (ed.) EuroPKI 2013. LNCS, vol. 8341, pp. 12–30. Springer, Heidelberg (2014)
13. Canard, S., Jambert, A.: On extended sanitizable signature schemes. In: Pieprzyk, J. (ed.) CT-RSA 2010. LNCS, vol. 5985, pp. 179–194. Springer, Heidelberg (2010)
14. Canard, S., Jambert, A., Lescuyer, R.: Sanitizable signatures with several signers and sanitizers. In: Mitrokotsa, A., Vaudenay, S. (eds.) AFRICACRYPT 2012. LNCS, vol. 7374, pp. 35–52. Springer, Heidelberg (2012)
15. Canard, S., Laguillaumie, F., Milhau, M.: Trapdoor sanitizable signatures and their application to content protection. In: Bellovin, S.M., Gennaro, R., Keromytis, A.D., Yung, M. (eds.) ACNS 2008. LNCS, vol. 5037, pp. 258–276. Springer, Heidelberg (2008)
16. Cavoukian, A., Polonetsky, J., Wolf, C.: Smartprivacy for the smart grid: embedding privacy into the design of electricity conservation. Identity in the Information Society 3(2), 275–294 (2010)
17. Chang, E.-C., Lim, C.L., Xu, J.: Short Redactable Signatures Using Random Trees. In: Fischlin, M. (ed.) CT-RSA 2009. LNCS, vol. 5473, pp. 133–147. Springer, Heidelberg (2009)
18. de Meer, H., Pöhls, H.C., Posegga, J., Samelin, K.: Scope of security properties of sanitizable signatures revisited. In: ARES, pp. 188–197 (2013)
19. Freeman, D.M.: Improved security for linearly homomorphic signatures: A generic framework. In: Fischlin, M., Buchmann, J., Manulis, M. (eds.) PKC 2012. LNCS, vol. 7293, pp. 697–714. Springer, Heidelberg (2012)
20. Goldwasser, S., Micali, S., Rivest, R.L.: A Digital Signature Scheme Secure Against Adaptive Chosen-Message Attacks. SIAM Journal on Computing 17, 281–308 (1988)
21. Gong, J., Qian, H., Zhou, Y.: Fully-secure and practical sanitizable signatures. In: Lai, X., Yung, M., Lin, D. (eds.) Inscrypt 2010. LNCS, vol. 6584, pp. 300–317. Springer, Heidelberg (2011)
22. Haber, S., Hatano, Y., Honda, Y., Horne, W.G., Miyazaki, K., Sander, T., Tezoku, S., Yao, D.: Efficient signature schemes supporting redaction, pseudonymization, and data deidentification. In: ASIACCS, pp. 353–362 (2008)
23. Hanser, C., Slamanig, D.: Blank digital signatures. In: AsiaCCS, pp. 95–106. ACM (2013)
24. Izu, T., Izumi, M., Kunihiro, N., Ohta, K.: Yet another sanitizable and deletable signatures. In: AINA, pp. 574–579 (2011)
25. Izu, T., Kunihiro, N., Ohta, K., Sano, M., Takenaka, M.: Sanitizable and deletable signature. In: Chung, K.-I., Sohn, K., Yung, M. (eds.) WISA 2008. LNCS, vol. 5379, pp. 130–144. Springer, Heidelberg (2009)
26. Johnson, R., Molnar, D., Song, D., Wagner, D.: Homomorphic signature schemes. In: Preneel, B. (ed.) CT-RSA 2002. LNCS, vol. 2271, pp. 244–262. Springer, Heidelberg (2002)
27. Klonowski, M., Lauks, A.: Extended Sanitizable Signatures. In: Rhee, M.S., Lee, B. (eds.) ICISC 2006. LNCS, vol. 4296, pp. 343–355. Springer, Heidelberg (2006)
28. Kundu, A., Bertino, E.: Structural Signatures for Tree Data Structures. In: Proc. of PVLDB 2008, New Zealand. ACM (2008)
29. Kundu, A., Bertino, E.: How to authenticate graphs without leaking. In: EDBT, pp. 609–620 (2010)
30. Kundu, A., Bertino, E.: Privacy-preserving authentication of trees and graphs. Intl. J. of Inf. Sec., 1–28 (2013)

31. Lim, S., Lee, E., Park, C.-M.: A short redactable signature scheme using pairing. Sec. and Comm. Netw. 5(5), 523–534 (2012)
32. Miyazaki, K., Hanaoka, G., Imai, H.: Digitally signed document sanitizing scheme based on bilinear maps. In: ASIACCS 2006, pp. 343–354. ACM, New York (2006)
33. Miyazaki, K., Iwamura, M., Matsumoto, T., Sasaki, R., Yoshiura, H., Tezuka, S., Imai, H.: Digitally Signed Document Sanitizing Scheme with Disclosure Condition Control. IEICE Transactions 88-A(1), 239–246 (2005)
34. Miyazaki, K., Susaki, S., Iwamura, M., Matsumoto, T., Sasaki, R., Yoshiura, H.: Digital documents sanitizing problem. Technical report, IEICE (2003)
35. Pöhls, H.C., Peters, S., Samelin, K., Posegga, J., de Meer, H.: Malleable signatures for resource constrained platforms. In: Cavallaro, L., Gollmann, D. (eds.) WISTP 2013. LNCS, vol. 7886, pp. 18–33. Springer, Heidelberg (2013)
36. Pöhls, H.C., Samelin, K., Posegga, J.: Sanitizable Signatures in XML Signature - Performance, Mixing Properties, and Revisiting the Property of Transparency. In: Lopez, J., Tsudik, G. (eds.) ACNS 2011. LNCS, vol. 6715, pp. 166–182. Springer, Heidelberg (2011)
37. Samelin, K., Pöhls, H.C., Bilzhause, A., Posegga, J., de Meer, H.: On Structural Signatures for Tree Data Structures. In: Bao, F., Samarati, P., Zhou, J. (eds.) ACNS 2012. LNCS, vol. 7341, pp. 171–187. Springer, Heidelberg (2012)
38. Samelin, K., Pöhls, H.C., Bilzhause, A., Posegga, J., de Meer, H.: Redactable signatures for independent removal of structure and content. In: Ryan, M.D., Smyth, B., Wang, G. (eds.) ISPEC 2012. LNCS, vol. 7232, pp. 17–33. Springer, Heidelberg (2012)
39. Slamanig, D., Rass, S.: Generalizations and extensions of redactable signatures with applications to electronic healthcare. In: De Decker, B., Schaumüller-Bichl, I. (eds.) CMS 2010. LNCS, vol. 6109, pp. 201–213. Springer, Heidelberg (2010)
40. Steinfeld, R., Bull, L., Zheng, Y.: Content extraction signatures. In: Kim, K.-C. (ed.) ICISC 2001. LNCS, vol. 2288, pp. 285–304. Springer, Heidelberg (2002)
41. Yuen, T.H., Susilo, W., Liu, J.K., Mu, Y.: Sanitizable signatures revisited. In: Franklin, M.K., Hui, L.C.K., Wong, D.S. (eds.) CANS 2008. LNCS, vol. 5339, pp. 80–97. Springer, Heidelberg (2008)
42. Yum, D.H., Seo, J.W., Lee, P.J.: Trapdoor sanitizable signatures made easy. In: Zhou, J., Yung, M. (eds.) ACNS 2010. LNCS, vol. 6123, pp. 53–68. Springer, Heidelberg (2010)

A   Requirements to Transform a $\mathcal{SSS}$ into a $\mathcal{RSS}$

In this section, we show that standard private $\mathcal{SSS}$s are not enough to build weakly private $\mathcal{RSS}$s. Moreover, we prove that weak blockwise non-interactive public accountability is required to build an unforgeable $\mathcal{RSS}$. To formally express this intuitive goals, we need Theorems 5 and 6.

Theorem 5 (Any non strongly private $\mathcal{SSS}$ results in a non-weakly private $\mathcal{RSS}$). *If the transformed $\mathcal{SSS}$ is not strongly private, the resulting $\mathcal{RSS}$ is not weakly private.*

Proof. Let $\mathcal{A}$ be an adversary winning the strong privacy game as defined in Fig. 5. We can then construct an adversary $\mathcal{B}$, which wins the weak privacy game as defined in Fig. 6, using $\mathcal{A}$ as a black-box:

1. $\mathcal{B}$ receives the following keys from the challenger: $pk_{\mathrm{san}}, sk_{\mathrm{san}}, pk_{\mathrm{sig}}$ and forwards them to $\mathcal{A}$

2. $\mathcal{B}$ simulates the signing oracle using the oracle provided
3. Eventually, $\mathcal{A}$ returns its guess b^*
4. $\mathcal{B}$ outputs b^* as its own guess

Following the definitions, the success probability of $\mathcal{B}$ equals the one of $\mathcal{A}$. This proves the theorem.

Theorem 6 (No Transform can Result in a Strongly Private $\mathcal{RSS}$). *There exists no algorithm which transforms a secure $\mathcal{SSS}$ into a strongly private $\mathcal{RSS}$.*

Proof. Once again, every meaningful $\mathcal{SSS}$ must be immutable, which implies weak immutability due to Th. 2. Hence, we do not make any statements about schemes not weakly immutable. We show that any transform $\mathcal{T}$ achieving this property uses a $\mathcal{SSS'}$ which is not weakly immutable. Let $\mathcal{RSS'}$ denote the resulting $\mathcal{RSS}$. We can then derive an algorithm which uses $\mathcal{RSS'}$ to break the weak immutability requirement of the underlying $\mathcal{SSS}$ in the following way:

1. The challenger generates the two key pairs of the $\mathcal{SSS}$. It passes all keys but sk_{sig} to $\mathcal{A}$
2. $\mathcal{A}$ transforms the $\mathcal{SSS}$ into $\mathcal{RSS'}$ given the transform $\mathcal{T}$
3. $\mathcal{A}$ calls the oracle $\mathcal{SSS}$.Sign with a message $m = (1, 2)$
4. $\mathcal{A}$ calls $\mathcal{RSS'}$.Redact with $\text{MOD} = (1)$
5. If the resulting signature σ does not verify, abort
6. $\mathcal{A}$ outputs $(m', \sigma_{\mathcal{SSS}})$ of the underlying $\mathcal{SSS}$

As $\ell_m \neq \ell\text{MOD}(m)$, $(\text{MOD}(m), \sigma_{\mathcal{SSS}})$ breaks the weak immutability requirement of the $\mathcal{SSS}$. Moreover, as hiding redacted parts of a message is essential for strong privacy, no algorithm exists, which transforms a weakly immutable $\mathcal{SSS}$ into a strongly private $\mathcal{RSS}$, as ADM needs to be correctly recoverable. This proves the theorem. This concrete example is possible, as we only use required behavior.

Theorem 7 (Weak Blockwise Non-Interactive Public Accountability is Required for any Transform $\mathcal{T}$). *For any transformation algorithm $\mathcal{T}$, the utilized $\mathcal{SSS}$ must be weakly blockwise non-interactive publicly accountable to result in an unforgeable $\mathcal{RSS}$.*

Proof. Let $\mathcal{RSS'}$ be the resulting $\mathcal{RSS}$ from the given $\mathcal{SSS}$. Perform the following steps to show that the used $\mathcal{SSS}$ is not weakly blockwise non-interactive publicly accountable. In particular, let $\mathcal{A}$ be an adversary winning the unforgeability game, which is used by $\mathcal{B}$ to break the weak blockwise non-interactive public accountability of the used $\mathcal{SSS}$.

1. The challenger generates the two key pairs of the $\mathcal{SSS}$. It passes all keys but sk_{sig} to $\mathcal{B}$
2. $\mathcal{B}$ forwards all received keys to $\mathcal{A}$
3. $\mathcal{A}$ transforms the $\mathcal{SSS}$ into $\mathcal{RSS'}$ given the transform $\mathcal{T}$

4. Any calls to the signing oracle by $\mathcal{A}$ are answered genuinely by $\mathcal{B}$ using its own signing oracle
5. Eventually, $\mathcal{A}$ returns a tuple (m, σ_{RSS}) to $\mathcal{B}$
6. If the resulting signature does not verify or does not win the unforgeability game, $\mathcal{A}$ and therefore also $\mathcal{B}$ abort
7. $\mathcal{B}$ outputs the underlying message-signature pair $(m', \sigma_{SSS'})$

Following Fig. 3, $(m', \sigma_{SSS'})$ breaks the weak blockwise non-interactive public accountability requirement of the SSS, as there exists a block, which has not been signed by the signer, while the signer is accused by Detect. Moreover, the success probabilities are equal. The contrary, i.e., if the SSS used is not weakly blockwise non-interactive publicly accountable, the proof is similar. To achieve the correctness requirements, our accountability definition must hold blockwise.

Theorem 8 (No Unforgeable RSS can be Transformed into an SSS). *There exists no transform $\mathcal{T}$, which converts an unforgeable RSS into an unforgeable SSS.*

Proof. Let SSS' be the resulting SSS. Now perform the following steps to extract a valid forgery of the underlying RSS:

1. The challenger generates a key pair for an RSS. It passes pk to $\mathcal{A}$.
2. $\mathcal{A}$ transforms RSS into SSS' given the transform $\mathcal{T}$
3. $\mathcal{A}$ calls the oracle RSS.Sign with a message $m = (1, 2)$ and simulates SSS'.Sign with $\text{ADM} = (1)$
4. $\mathcal{A}$ calls SSS'.Sanit with $\text{MOD} = (1, a)$, $a \in_R \{0, 1\}^\lambda$.
5. If the resulting signature does not verify, abort
6. Output the resulting signature σ_{RSS} of the underlying RSS

As $(a, 2) \notin \text{span}_{\not\models}(m)$, $((a, 2), \sigma_{RSS})$ is a valid forgery of the underlying RSS. Note, this concrete counterexample is possible, as only required behavior is used.

B   Proofs of Theorem 3 and 4

Th. 3: There exists an SSS which is private, but not strongly private.

Proof. We do so by modifying an arbitrary existing strongly private SSS. Let $SSS = (\text{KGen}_{sig}, \text{KGen}_{san}, \text{Sign}, \text{Sanit}, \text{Verify}, \text{Proof}, \text{Judge})$ be an arbitrary private SSS. We alter the scheme as follows:

- $\text{KGen}'_{sig} := \text{KGen}_{sig}$
- $\text{KGen}'_{san} := \text{KGen}_{san}$, while an additional key pair for a IND-CCA2-secure encryption scheme ENC is generated.
- Sign' is the same as Sign, but it appends the encryption e of a digest of original message to the final signature, i.e., $\sigma' = (\sigma, e)$, where $e \leftarrow \text{ENC}(pk_{san}, \mathcal{H}(m))$ and $\mathcal{H}$ some cryptographic hash-function.
- Sanit' is the same as Sanit, while it first removes the encrypted digest from the signature, appending it to the resulting signature.

– Verify', Proof' and Judge' work the same as their original counterparts, but removing the trailing e from the signature before proceeding.

Clearly, a sanitizer holding the corresponding secret key for ENC, can distinguish between *messages* generated by the signer and the sanitizer using the information contained in the *signature* σ. Without sk_{san}, this information remains hidden due to the IND-CCA2 encryption.

Th. 4: Our Construction is Secure. We have to show that the resulting $\mathcal{RSS}$ is unforgeable and weakly private, but not strongly.

Proof. We prove each property on its own.

I) **Unforgeability.** Let $\mathcal{A}$ be an algorithm breaking the unforgeability of the resulting $\mathcal{RSS}$. We can then construct an algorithm $\mathcal{B}$ which breaks the weak blockwise non-interactive public accountability of the utilized $\mathcal{SSS}$. To do so, $\mathcal{B}$ simulates $\mathcal{A}$'s environment in the following way:
   1. $\mathcal{B}$ receives $pk_{san}, sk_{san}, pk_{sig}$ and forwards them to $\mathcal{A}$
   2. $\mathcal{B}$ forwards every query to its own signing oracle
   3. Eventually, $\mathcal{A}$ outputs a tuple (m^*, σ^*)
   4. If (m^*, σ^*) does not verify or is trivial, abort
   5. $\mathcal{B}$ outputs (m^*, σ^*)

   m cannot be derived from any queried message, with the exception of $m[i] = \square$ for any index i. Hence, $\exists i : m[i] \neq \square$, which has not been signed by the signer. The accepting verification requires that $\mathtt{Sig} = \mathtt{Detect}(1^\lambda, m^*, \sigma^*, pk_{sig}, pk_{san})$. Therefore, (m^*, σ^*) breaks the weak blockwise non-interactive publicly accountability. The success probability of $\mathcal{B}$ equals the one of $\mathcal{A}$.

II) **Weak Privacy.** To show that our scheme is weakly private, we only need to show that an adversary $\mathcal{A}$ cannot derive information about the prior content of a contained block $m[i]$, as $\square$ is considered part of the resulting message m' and all other modifications result in a forgeable $\mathcal{RSS}$. Let $\mathcal{A}$ winning the weak privacy game. We can then construct an adversary $\mathcal{B}$ which breaks the strong privacy game in the following way:
   1. $\mathcal{B}$ receives $pk_{san}, sk_{san}, pk_{sig}$ and forwards them to $\mathcal{A}$
   2. $\mathcal{B}$ forwards every query to its own oracles
   3. Eventually, $\mathcal{A}$ outputs its guess b^*

   $\mathcal{B}$ outputs b^* as its own guess. The oracle requires that $\mathrm{MOD}_1(m_1) = \mathrm{MOD}(m_2)$, disregarding $\square$. Note, the messages are the same. Hence, the success probability of $\mathcal{B}$ is the same as $\mathcal{A}$'s. This proves the theorem.

III) **No Strong Privacy.** Due to the above, we already know that our scheme is weakly private. Hence, it remains to show that it is not strongly private. As a redaction leaves a *visible* special symbol, i.e., $\square$, an adversary can win the *strong* privacy game in the following way: Generate two messages m_0, m_1, where $m_1 = (m_0, 1)$. Hence, $\ell_0 < \ell_1$, while m_0 is a prefix of m_1. Afterwards, it requests that $m_1[\ell_1]$ is redacted, i.e., $\mathrm{MOD}_1 = (\ell_1)$ and $\mathrm{MOD}_0 = ()$. Hence, if the oracle chooses $b = 0$, it will output $m_2 = m_0$ and for $b = 1$, $m_2 = (m_1, \square)$. Hence, the adversary wins the game, as $(m_1, \square) \neq m_0$.

Idea: Towards a Working Fully Homomorphic Crypto-processor
Practice and the Secret Computer

Peter T. Breuer[1] and Jonathan P. Bowen[2,*]

[1] Department of Computer Science, University of Birmingham, UK
ptb@cs.bham.ac.uk
[2] School of Computing, Telecommunications and Networks,
Birmingham City University, UK
jonathan.bowen@bcu.ac.uk

Abstract. A KPU is a replacement for a standard RISC processor that natively runs encrypted machine code on encrypted data in registers and memory – a 'general-purpose crypto-processor', in other words. It works because the processor's arithmetic is customised to make the chosen encryption into a mathematical homomorphism, resulting in what is called a 'fully-homomorphic encryption' design. This paper discusses the problems and solutions associated with implementing a KPU in hardware.

1 Introduction

A KPU ('**K**rypto-**P**rocessor **U**nit') is a simple general purpose processor and processor architecture that works on data in encrypted form. To input data into the machine, the owner prepares it in encrypted form and receives encrypted data back. In theory, a KPU need never decrypt, even as it places encrypted data in memory and registers and runs the encrypted program. That makes it of interest for cloud computation, and also the reverse situation, where, for example, a bank wishes to securely devolve responsibility for transactions on individual bank accounts to personal chips held by the untrusted account owners.

The mathematics relates a KPU to the science of fully-homomorphic encryption, first introduced as privacy homomorphisms in [10]. Well-known encryptions such as RSA private/public key cryptography [11] exhibit partial homomorphism, in the case of RSA with respect to multiplication, in that $RSA(a)*RSA(b)$ mod $m = RSA(a*b)$, where RSA stands for the encryption and m is its associated (large) arithmetic modulus. An encryption that supports homomorphism both with respect to addition and to multiplication is said to be a fully homomorphic encryption (FHE). The operations on encrypted data corresponding to multiplication and addition on the unencrypted data need not be as simple as multiplication or addition in the general case, but Gentry [6] constructed a FHE in which those operations, while complex, do not compromise the encryption

---

[*] Jonathan Bowen acknowledges the support of Museophile Limited.

J. Jürjens, F. Piessens, and N. Bielova (Eds.): ESSoS 2014, LNCS 8364, pp. 131–140, 2014.

and thus may still be carried out by untrusted parties. In a KPU, the corresponding operations are embedded in the hardware. A KPU can be handed out to an untrusted party just as the software algorithms for the operations that work on FHE-encrypted data can be handed out. However, a KPU design allows more flexibility in the choice of the underlying encryption, and engineering tradeoffs come to the fore. One may entertain, for example, the possibility of operations that would reveal the encryption if they were exposed, but which are implemented in hardware that physically secures them, as for SmartCards [8].

In terms of its construction, a KPU is a processor in which the standard arithmetic logic unit (ALU) has been replaced by a different design satisfying certain special properties. A standard processor is the trivial case. A KPU is described in mathematical terms in [1], where it is shown that, whatever the detail of the implementation, provided the modified ALU satisfies the required properties then a KPU operates *correctly*, in that the machine states that obtain during the execution of an encrypted program are encryptions of the states that would result in an ordinary RISC [9] processor running the unencrypted program. A RISC design is a convenient point of departure for the proof, because of its simplicity, but it is equally possible to build a KPU by starting from another class of modern von Neumann processor.

The modified ALU in a KPU, instead of $1 + 1 = 2$, does $6769875\#6769875 = 87997001$ (for example), those numbers being encodings of 1 and 2 under the encryption. In its most general form, this is a homomorphism statement, and the 'special properties' of the modified ALU alluded to above are the requirements that its operations, functions and relations be homomorphic images of the standard operations.

Counter-intuitively for those who appreciate that the slightest change inside the processor may snowball, the grossly changed arithmetic results in states that are 'correct', but encrypted. One can liken it to changing from speaking 'English' in a CPU to speaking 'Chinese' in the KPU, with the added difficulty of very many 'Chinese' words for every 'English' word. The detail in a practical KPU design is merely aimed at avoiding design features that may inadvertently sabotage this principle. It is important, for example, to separate the circuitry that does arithmetic on program addresses from that which does arithmetic on data, or the encrypted '+1' on the address at most every tick of the clock would leak significant information, as well as compromise program loading and caching localities. In consequence, while data addresses and contents and program instructions are encrypted in a KPU, program addresses should be encrypted differently and may not be encrypted at all. Running programs must be written to keep the two kinds apart, so that encrypted values are acted on by instructions that expect encryption, and unencrypted values are acted on by instructions that do not expect encryption [2].

The idea behind KPU design is 'problem reduction'. It reduces the problem of encrypted general purpose computing to the lower order problem of constructing an appropriately modified ALU. In principle a very large lookup table suffices to drive the ALU, but replacing every gate in the standard ALU with an 'encrypted version' of the same gate also works. Between those extremes lie many other possibilities, which will be explored in this paper.

Speed is not a primary concern — IBM's FHE implementations take on the order of seconds per single bit operation on a vector mainframe [7], though this shows signs of being improved by means of special techniques on GPU-based hardware [12] — and there are many other factors to consider. There is, for example, a *hardware aliasing problem*, which arises because ciphers are one–many and thus many different ('Chinese') encryptions of a single ('English') memory address may crop up and be used during the execution of a program. Since all the different aliases designate physically different locations in memory, a working program for a KPU must be written to access only one of them, which means every address must be calculated the same way at every use [3].

This paper focuses on two areas in working KPU design in particular. Section 2 discusses hardware options and Section 3 discusses encoding strategies.

2 Word Size and Hardware Design

The first issue in processor design is 'how big is a data word', the physical extent of the standard unit of data. 'Large' means the processor needs long registers and many traces and wide busses to carry the data internally, which is costly. The quick answer here is that nobody yet really knows what word size is best, because different design approaches indicate different solutions.

In a KPU, the data word size is the encryption block size, the size of a unit of encrypted data. For strong encryption, about 128 bits is reasonable for many of the common ciphers in the medium to long term, whereas 64 bits is borderline, but sufficient for real time protection, supposing key size the same as encryption block size. Neither is technically impossible, but 64 bits would be very favourable from the manufacturer's point of view as the associated technology is already in use. Fewer bits would be even better from that perspective, however. The trade-off is small size (low cost, low power) against greater security.

How many plaintext bits does a 64-bit encrypted data word contain? It can be anywhere from 1 to close to 64, leaving room for padding bits that make the encryption one–many overall; 32 plaintext bits and 32 padding bits would result in 2^{32} different encodings of each 32-bit plaintext number. The numbers are significant because if an attacker guesses the encryption for 1 and also guesses which operation is the '+' in the ALU, then, given access, the attacker can generate the encryption of 2 via $1 + 1$, of 3 via $2 + 1$, etc, and thus obtain a complete codebook. Many encodings for each plaintext number imply many codebooks and only relatively few encountered in any program run. That is reassuring because, in general, there is no mathematical analysis available of the security of arithmetic in a KPU. That is also the case for white-box access, but note that the example of the encrypted arithmetic in Gentry's software solution [5] shows that it is not a priori unsafe to permit unfettered access to hardware. To make the discussion concrete, five designs are set out below.

1. Embedded codecs: We may create an encrypted ALU by placing codecs on inputs and outputs of a standard ALU, as in Fig. 1. The codecs (D, C) contain keys that must be transferred securely into the hardware, perhaps via a Diffie-Hellman protocol [4], and they must be invulnerable to electronic probes. That is within the capabilities of SmartCard manufacturers today. The key should be volatile, so it does not survive disconnection, and the hardware's internal traces protected by overlying circuit elements. A 64-bit (encrypted) word is inherently feasible, but stripping out and replacing 32 bits of padding requires extra hardware. The simplest implementation has the data bits in the middle of the word and routes them to a 32-bit ALU. The output padding is generated by a separate unit (P); it can multiply input paddings and take the middle 32 bits ('mid-out' hash), folding in extra randomness as desired.

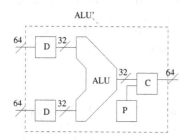

Fig. 1. Building an encrypted ALU using codecs (D, C) with embedded keys, and a padding unit (P). 64-bit inputs at left, 64-bit outputs at right.

One problem with this kind of design, apart from potential vulnerabilities with harbouring keys, is that the speed of the codecs limits the speed of the processor. Some encryptions when done in hardware can run at a few hundred MHz (CuBox run an ARM v6-based chip doing AES encryption clocked at 800MHz).

Nevertheless, the design is easily realised with present-day silicon technologies, requiring no radical innovations, and represents the most likely initial implementation technique. A small company with processor expertise can already produce chip wafers based on this idea.

2. Lookup tables: Fig. 2 shows a 16-bit 'black-box' ALU design. It contains tables for encrypted arithmetic on two 16-bit encrypted inputs, producing one 16-bit encrypted output. Leaving security questions aside (16-bit cipherspaces are easily searched, but that is not the end of the story), each binary operator requires tables occupying $(2^{16})^2 \times 16 = 2^{36}$ bits, or 8GB. That is too large to go in-processor at present, but it can reside in RAM. Every arithmetic operation must access the tables, which limits speeds to 200MHz to 400MHz, in practice. But we expect advances in technology to make the numbers feasible in a few years, returning focus to the security question here.

This solution focuses on encrypted arithmetic tables as the encryption 'key'. Those 8GB lumps of data in RAM need to be supplied, probably over the Internet, at relatively frequent intervals as different configurations are adopted for different encryptions, but sending them beforehand, or in parallel while computation proceeds, is an option. The transfer need not be in public view, but if it

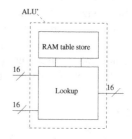

Fig. 2. Building an encrypted ALU as a 'black-box' lookup table requires large amounts of on-board RAM. 16-bit inputs at left, 16-bit outputs at right.

is, then can an attacker work out an encoding given the full tables, together with observations of what computations are done in practice?

The answer is formally 'no' (one may place two and in practice 'very many' encodings simultaneously in the same tables; see Section 3), but observations of a running CPU may aid the attacker. If this solution is secure, then it can also be implemented in software, leading to the safe running of a KPU in simulation.

The problematic aspect of this solution is the short 16-bit input and output sizes. If, say, 8 bits of that is padding, it only leaves room for 8 bits of data beneath the encryption. While 8-bit computation is feasible (and from 4 to 16 times as slow as 32-bit computation), it is disadvantaged in security terms because 32-bit calculations take several cycles, and the carry in to the second cycle will be highly constrained and yet encrypted in the same way as the other inputs, which makes decoding relatively easy.

On the plus side, however, is that algebraic attacks using the ring structure of addition and multiplication under the homomorphism do not work. One might look for, say, 3 as one of the highest-order elements of the tables under self-multiplication (it should take or 2^{16} or 16 self-multiplications to get back to 3 again, depending how one counts), but that approach is scotched because the padding makes the result still look different from the original.

3. Modular design: Putting several ALUs in parallel in the hardware allows the lookup table solution to be 'ramped up' to deliver 32-bit computation in one cycle in hardware, as shown in Fig. 3 for addition. The individual encrypted adders are 16-bit for a total of 64 bits of input and output, but the encryption is different in each group of 16 bits. The encryption varies again on each of the carry outputs and inputs. One may imagine that between each adder an arbitrary extra encryption has been applied via codecs D_i, C_i, shown in dotted lines, but in reality these are folded into the tables.

In hardware, no internal connections are exposed, but this solution is problematic in software. Can the units in solid lines be safely stored as lookup tables in full public view? The answer is formally unknown.

One may embed at least two different ciphers in one 3-bit encrypted arithmetic table, encoding just one data bit (the technique is explained in Section 3). The maximum number M of ciphers is much higher than two, but there are too many configurations to establish M exactly (a 3-bit table for one arithmetic operation has 64 entries, each 3 bits, thus $8^{64} = 2^{192}$ tables to explore). The significance of that may be seen via an analogy: Suppose that the English word 'mouse' is also the Chinese word for 'sunshine', with similar overlaps for all English and Chinese words. The situation here is then that an observer cannot decide if an observed computation is an 'English' conversation about pests or a 'Chinese' conversation about weather. The mathematics makes the 'grammar' (the arithmetic) look right both ways.

The layout of Fig. 3 may be adapted for 32 3-bit units, each encoding one bit of data. Then 32-bit computation is implemented with 96-bit encrypted words and an attacker with full access must explore M^{32} valid decodings, assuming it is already known which decodings are valid for those tables (there are only

$8!/2! = 20160$ each to check in the 3-bit case). The tables are small and may be placed in-processor. Alternatively, 21-bit computation (via 21 3-bit units in the layout of Fig. 3) can be fitted in 63-bit encrypted words, and 24-bit computation in 64-bit encrypted words is feasible using trits and base-6 digitisation.

The lemma to remember here is that the arithmetic tables for coded values do not expose the coding, when padding makes the coding 1-to-many.

We will elaborate the approach in §5 below. First consider another approach that at first also looks unlikely.

4. Gate-level encryption: One may replace every single (1-bit) trace in an ALU by a set of 3 or 8 or 16, etc., traces carrying respectively a 3- or 8- or 16-bit encryption of the single bit; every OR gate is replaced by a corresponding 'encrypted-OR gate', possibly table-driven. If we consider the units of Fig. 3, then each of them may be implemented via a network of such 'encrypted gates'.

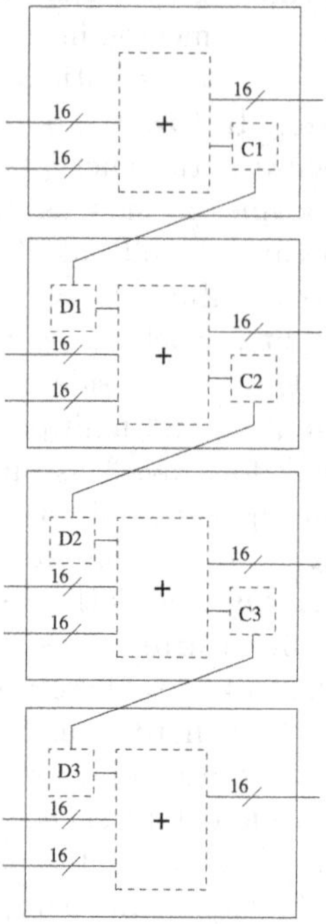

This reduces the design problem to dealing with just one bit of data, encrypted in as many bits as may be advisable for security. Fig. 4 shows how a 16-bit encrypted one-bit half-adder may be implemented like this. Instead of one table for addition and another for multiplication, etc., there is one table for 1-bit AND, one for 1-bit OR, and one for 1-bit inversion, but, in theory, just one table, for 1-bit NAND, will suffice. So storage requirements are 'only' one 8GB table for each 16-bit encryption used.

An entire 16-bit encrypted ALU can be built in this way, using just one encryption, but it requires 16×16 input and output traces, i.e., encrypted words of 256 bits. But there is a problem: how to access the arithmetic tables simultaneously for all the gates that need to. In practice, with today's technology, one cannot. Even so, ALUs tend to be built so that calculation propagates across them in systolic fashion. If the construction is at

Fig. 3. Encrypted adder built from smaller units, with 4×16-bit inputs at left, 4×16-bit output at right, and distinct encodings in each unit.

worst n gates wide by m deep (m determines the latency), then the calculation may in principle be pipelined using just n gates and n tables organised into m stages. Each stage of the calculation takes time equal to one table lookup, thus a complete arithmetic calculation should take time equal to m table lookups. But one complete calculation will exit the pipeline at intervals of one lookup, so the throughput is normal. Pipelined ALUs (for floating point arithmetic) are common in processor technology.

5. Hybrids: We now revisit the modular design of Fig. 3. That has relatively weak 4×16-bit security in the configuration shown, but imagine an AES codec that *decrypts* a 64-bit ciphertext to a 64-bit plaintext, but which does so in hardware. AES works via addition and multiplication operations on 16-bit segments. Apply the process that built Fig. 4 to the codec, operation by operation.

Imagine the ith 16-bit segment of plaintext output as followed by a new encoder E_i that produces 16 bits of encrypted output. Pass E_i to the input side of the adjacent internal AES operation G, replacing it with an encrypted operation G', such that $E_i \cdot G' = G \cdot E_i$. Repeat, passing the encoders E_i from output to input side of each successive layer of internal AES operations in turn.

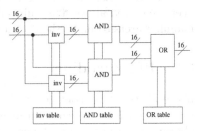

The process ends in a design analogous to Fig. 4, shown at left in Fig. 5 as 'D'. In it, every 16-bit operation G in the original AES decoder has been replaced by an 'encrypted version' G' working on 16-bit encrypted words. Internal changes between encryptions E_i and E_j are notionally handled by extra codecs Di and Cj shown in Fig. 5, although in reality these are folded into the tables that drive the different encrypted operations. What does this bizarre construction do?

Fig. 4. An adder built from encrypted gates. 16-bit inputs at left, 16-bit output at right.

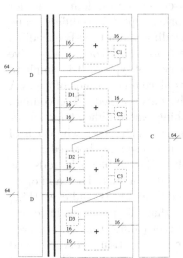

Fig. 5. Increasing encryption security. 64-bit inputs at left, 64-bit output at right.

The answer is that it takes as input $4 \times 16 = 64$ bits in which the ith group of 16 is an encoding under E_i of the ith 16-bit segment of the AES encoding of a 64-bit plaintext number x. The output is a 64-bit word y in which the ith group of 16 bits is the encoding under E_i of the ith 16-bit segment of x. In other words, it decodes doubly encrypted data to singly encrypted data that is suitable as input to the design of Fig. 3. The AES key is kept encrypted under the E_i in the internals of the modified decoder.

After decoding by D, doubly encrypted data is suitable for handling by the encrypted arithmetic structure of Fig. 3, shown in the centre of Fig. 5. The output from that may be recoded again using a modified AES encoder, labelled C in Fig. 5, producing doubly encrypted data. The AES keys are stored encrypted, and the ALU of Fig. 5 does 'doubly encrypted arithmetic'. An untrusted party with the encrypted AES keys can decrypt doubly encrypted data to singly encrypted data, but no more.

What is the advantage of this construction? The input encoding is at least as safe as AES. The keys, even if uncovered, are themselves encrypted and do not serve to decrypt the input, or output, or any intermediate. But the number of bits in the construction is most significant: it is just 64 bits. If every trace and gate had been 'encrypted', as in §4, it would have been 32×64 bits.

So what is the right word size? Every size from 3, 8, 16, 64, 256, 512 bits has been suggested above. It requires detailed simulation and negotiation with chip manufacturers to make the decision in practice.

3 ABC Encoding

If the KPU's ALU contains a division operator – or even if there is a division routine in software – then an attacker with sufficient access can nearly always obtain an encryption of 1 by computing x/x for any encrypted x that has been observed. Even if which operation is division is formally unknown, the choice of n (usually 64 at most in a conventional ALU) operations merely multiplies up the number of possibilities to be tried by n, which is not significant. And obtaining encrypted 1 gives encrypted 2 via $1+1$, etc, until a complete codebook is constructed. This attack has implications for the encodings used in a KPU.

We have remarked that padding under the encryption is justified by the need for many codebooks in order to confound this kind of attack, but there is also a separate 'defense by construction' that may be built into the system. It is to set the ALU so that $x \operatorname{op} x$, always gives a nonsensical or random result, for any operator. How then to *really* calculate $1/1$, for example? Two different and disjoint encodings are implemented, a type-A encoding and a type-B encoding, and A op B gives the right answer of type B, while A op A and B op B always give nonsense. Symmetrically, B op A gives the right answer of type A. It is simple to compile programs such that every operation takes place on operands of types A and B, or B and A, and the program runs correctly. This is called 'AB encoding'.

Unfortunately, AB encoding does not make things more difficult for an attacker with sufficient access. Doing the calculation $(x+y)/(y+x)$, where x is of type A and y of type B, still gives 1. An attacker may use any observed x, y.

An improvement called 'ABC encoding' resolves the problem. It adds one more disjoint encoding, a so-called type-C encoding, to the mix. The valid operations are now of type $A \operatorname{op} B = C$, $B \operatorname{op} C = A$ and $C \operatorname{op} A = B$, and everything else gives nonsense. Again, compiling programs so that all operations have correct typing is trivial. One may prove that an attacker cannot take an observed calculation x of type A, reshuffle its parts to obtain a calculation y of type B, and then compute x/y with ABC encoding, for a known constant result. The logic of ABC encoding does not allow it. A loophole, however, is that the attacker may not merely reshuffle parts, but also duplicate or eliminate some parts in a revised sum, to get a constant. The following calculation is valid in ABC encoding:

$$(x * y) + (y * (x * y)) = 0 \quad \mod 2$$

So an attacker with the encrypted arithmetic tables for two 1-bit plaintext operations can obtain an encoding of the 1-bit plaintext '0'. If the attacker does not know which operation is which, however, then there is no way of getting a constant result out (mod 2), and nothing can be gained in this way.

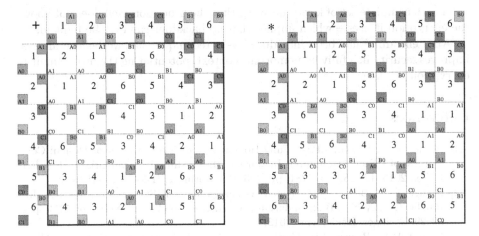

Fig. 6. Two different encodings simultaneously embedded in the same 6×6 tables for encrypted addition and multiplication mod 2, using ABC encoding. The 'lower left' A, B, C encryptions of 0,1 are 1,2;3,4;5,6 respectively. The 'upper right' encryptions are 2,1;6,5;3,4 respectively. Only AB=C, BC=A, CA=B gives valid results, such as A0+B1=C1, the rest are arbitrary.

ABC encoding trebles the size of the cipherspace required, and thus requires nine times as much storage space for arithmetic tables, as well as slightly increasing the number of bits required for an encrypted word. However, several different encryptions may be placed in the same tables simultaneously. Fig. 6 shows an example in the minimal possible size ABC tables: 6×6 in a cipherspace of size 6 for modulo 2 arithmetic.

In practice, 8 coding values would be used for 3 full bits of cipherspace, the redundancy permitting more overlaps. However, the upper limit for overlap as the cipherspace size increases is not known, nor is the number $M \geq 2$ of different encryptions that may be fitted in simultaneously. Despite the formal unknowns, we believe that ABC encodings do make it much more difficult to deduce the encryption from the encrypted arithmetic tables. An attacker may never hope to recognise a chance encoding that looks like $1 * 1 = 1$ under ABC rules, for example, because patterns of that form may never be constructed.

Fig. 6 proves that the tables do not determine the encryption uniquely.

4 Conclusion

The construction of the modified arithmetic logic unit in a KPU (a general purpose fully homomorphic crypto-processor architecture) has been discussed, with options ranging from monolithic lookup tables to gate-wise encryption. The objective is the implementation of PC-sized or smaller-sized computers that do all their work encrypted, with applications in many areas in the realm of secure computing. There is a close relation with work on fully-homomorphic computing, with hardware replacing the role of software algorithms. Some design options use no or an incomplete set of keys – meaning that the processor itself does not know the encryption it uses – which implies that no backdoor can ever be built in.

'ABC encoding' has been introduced as a technique that we believe always improves security in a fully homomorphic context, potentiating the use of smaller encryption block-sizes.

References

[1] Breuer, P.T., Bowen, J.P.: A Fully Homomorphic Crypto-Processor Design: Correctness of a Secret Computer. In: Jürjens, J., Livshits, B., Scandariato, R. (eds.) ESSoS 2013. LNCS, vol. 7781, pp. 123–138. Springer, Heidelberg (2013)

[2] Breuer, P.T., Bowen, J.P.: Typed Assembler for a RISC Crypto-Processor. In: Barthe, G., Livshits, B., Scandariato, R. (eds.) ESSoS 2012. LNCS, vol. 7159, pp. 22–29. Springer, Heidelberg (2012)

[3] Breuer, P.T., Bowen, J.P.: Certifying Machine Code Safe from Hardware Aliasing: RISC is not necessarily risky. In: Counsell, S., Núñez, M. (eds.) SEFM 2013 Collocated Workshops, LNCS, vol. 8368, Springer, Heidelberg (2014)

[4] Diffie, W., Hellman, M.: New directions in cryptography. IEEE Transactions on Information Theory 22(6), 644–654 (1976), doi:10.1109/TIT.1976.1055638.

[5] Gentry, C.: Computing arbitrary functions of encrypted data. Communications of the ACM 53(3), 97–105 (2010)

[6] Gentry, C.: Fully Homomorphic Encryption Using Ideal Lattices. In: Proc. 41st ACM Symposium on Theory of Computing (STOC), pp. 169–178. ACM (2009), doi:10.1145/1536414.1536440, ISBN: 978-1-60558-506-2

[7] Gentry, C., Halevi, S.: Implementing Gentry's fully-homomorphic encryption scheme. In: Paterson, K.G. (ed.) EUROCRYPT 2011. LNCS, vol. 6632, pp. 129–148. Springer, Heidelberg (2011)

[8] Kömmerling, O., Kuhn, M.G.: Design principles for Tamper-Resistant Smartcard Processors. In: Smartcard 1999, Chicago, Illinois, USA, May 10-11, pp. 9–20 (1999)

[9] Patterson, D.A.: Reduced Instruction Set Computers. Communications of the ACM 28(10), 8–21 (1985)

[10] Rivest, R.L., Adleman, L., Dertouzos, M.L.: On data banks and privacy homomorphisms. Foundations of Secure Computation 32(4), 169–180 (1978)

[11] Rivest, R.L., Shamir, A., Adleman, L.: A method for obtaining digital signatures and public-key cryptosystems. Communications of the ACM 21(2), 120–126 (1978)

[12] Wei, W., et al.: Accelerating Fully Homomorphic Encryption on GPUs. In: Proc. IEEE High Performance Extreme Computing Conference (2012)

Architectures for Inlining Security Monitors in Web Applications

Jonas Magazinius, Daniel Hedin, and Andrei Sabelfeld

Chalmers University of Technology, Gothenburg, Sweden

Abstract. Securing JavaScript in the browser is an open and challenging problem. Code from pervasive third-party JavaScript libraries exacerbates the problem because it is executed with the same privileges as the code that uses the libraries. An additional complication is that the different stakeholders have different interests in the security policies to be enforced in web applications. This paper focuses on securing JavaScript code by *inlining* security checks in the code before it is executed. We achieve great flexibility in the deployment options by considering security monitors implemented as security-enhanced JavaScript interpreters. We propose architectures for inlining security monitors for JavaScript: via browser extension, via web proxy, via suffix proxy (web service), and via integrator. Being parametric in the monitor itself, the architectures provide freedom in the choice of where the monitor is injected, allowing to serve the interests of the different stake holders: the users, code developers, code integrators, as well as the system and network administrators. We report on experiments that demonstrate successful deployment of a JavaScript information-flow monitor with the different architectures.

1 Introduction

JavaScript is at the heart of what defines the modern browsing experience on the web. JavaScript enables dynamic and interactive web pages. Glued together, JavaScript code from different sources provides a rich execution platform. Reliance on third-party code is pervasive [32], with the included code ranging from format validation snippets, to helper libraries such as jQuery, to helper services such as Google Analytics, and to fully-fledged services such as Google Maps and Yahoo! Maps.

Securing JavaScript Securing JavaScript in the browser is an open and challenging problem. Third-party code inclusion exacerbates the problem. The *same-origin policy (SOP)*, enforced by the modern browsers, allows free communication to the Internet origin of a given web page, while it places restrictions on communication to Internet domains outside the origin. However, once third-party code is included in a web page, it is executed with the same privileges as the code that uses the libraries. This gives rise to a number of attack possibilities that include location hijacking, behavioral tracking, leaking cookies, and sniffing browsing history [21].

Security policy stakeholders An additional complication is that the different stakeholders have different interests in the security policies to be enforced in web applications.

J. Jürjens, F. Piessens, and N. Bielova (Eds.): ESSoS 2014, LNCS 8364, pp. 141–160, 2014.

Users might demand stronger guarantees than those offered by SOP when it is not desired that sensitive information leaves the browser. This makes sense in popular web applications such as password-strength checkers and loan calculators. *Code developers* clearly have an interest in protecting the secrets associated with the web application. For example, they might allow access to the first-party cookie for code from third-party services, like Google (as needed for the proper functioning of such services as Google Analytics), but under the condition that no sensitive part of the cookie is leaked to the third party. *Code integrators* might have different levels of trust to the different integrated components, perhaps depending on the origin. It makes sense to invoke different protection mechanisms for different code that is integrated into the web application. For example, an e-commerce web site might include jQuery from a trusted web site without protection, while it might load advertisement scripts with protection turned on. Finally, *system and network administrators* also have a stake in the security goals. It is often desirable to configure the system and/or network so that certain users are protected to a larger extent or communication to certain web sites is restricted to a larger extent. For example, some Internet Service Providers, like Comcast, inject JavaScript into the users' web traffic but so far only to display browser notifications for sensitive alerts[1].

Secure inlining for JavaScript This paper proposes a novel approach to securing Java-Script in web applications in the presence of different stakeholders. We focus on securing JavaScript code by *inlining* security checks in the code before it is executed. A key feature of our approach is focusing on security monitors implemented, in JavaScript, as security-enhanced JavaScript interpreters. This, seemingly bold, approach achieves two-fold flexibility. First, having complete information about a given execution, security-enhanced JavaScript interpreters are able to enforce such fine-grained security policies as *information-flow security* [37]. Second, because the monitor/interpreter is itself written in JavaScript, we achieve great flexibility in the deployment options.

Architectures for inlining security monitors As our main contribution, we propose architectures for inlining security monitors for JavaScript: via browser extension, via web proxy, via suffix proxy (web service), and via integrator. While the code extension and proxy techniques themselves are well known, their application to security monitor deployment is novel. Being parametric in the monitor itself, the architectures provide freedom in the choice of where the monitor is injected, allowing to serve the interests of the different stake holders: users, code developers, code integrators, as well as system and network administrators.

We note that our approach is general: it applies to arbitrary security monitors, implemented as JavaScript interpreters. The Narcissus [13] project provides a baseline JavaScript interpreter written in JavaScript, an excellent starting point for supporting versatile security policies.

Our evaluation of the architectures explores the relative security considerations. When introducing reference monitoring, Anderson [3] identifies the following principles: (i) the monitor must be tamperproof *(monitor integrity)*, (ii) the monitor must be always invoked *(complete mediation)* [39], and (iii) the monitor must be small enough to be

---

[1] https://gist.github.com/ryankearney/4146814

subject to correctness analysis *(small trusted computing base (TCB))* [39,35]. Overall, the requirements often considered in the context of monitoring are that the monitor must enforce the desired security policy *(soundness)* and that the monitor is transparent to the applications *(transparency)*. Soundness is of higher priority than transparency in our setting. Our methods of deployment do not rely on transparency to provide security guarantees. Even if the application is able to detect that it is running in the monitor, this knowledge cannot be used to circumvent the monitor given that the monitor is sound (if this was possible the monitor is by definition not sound). Of course, if the application is able to detect that it is monitored it might chose to only expose benign behavior in order to escape detection. In either case, the user is protected from attacks. Note the relation of the soundness to Anderson's principles: while the principles do not automatically imply soundness, they facilitate establishing soundness. Transparency requirements are often in place for reference monitors to ensure that no new behaviors are added by monitors for any programs, and no behaviors are removed by monitors when the original program is secure.

Since the architectures are parametric in the actual monitor, we can draw on the properties of the monitor to guarantee the above requirements. It is essential for soundness and transparency that the monitor itself supports them. In our consideration of soundness for security, we assume the underlying monitors are sound (as natural to expect of such monitors). This implies that dealing with such features as dynamic code evaluation in JavaScript is already covered by the monitors. We note that monitor integrity, and complete mediation are particularly important in our security considerations because they are crucially dependent on the choice of the architecture. Our security considerations for the architectures are of general nature because of the generality of the security policies we allow.

Roadmap We study the relative pros and cons of the architectures. The goal of the study is not to identify a one-fits-all solution but to highlight the benefits and drawbacks for the different stakeholders. With this goal, we arrive at a roadmap to be used by the stakeholders when deciding on what architecture to deploy.

Instantiation To illustrate the usefulness of the approach, we present an instantiation of the architectures to enforce secure information flow in JavaScript. Information-flow control for JavaScript allows tracking fine-grained security policies for web applications. Typically, information sources and sinks are given sensitivity labels, for example, corresponding to the different Internet origins. Information-flow control prevents *explicit flows*, via direct leaks by assignment commands, as well as *implicit flows* via the control flow in the program.

Our focus on information flow is justified by the nature of the JavaScript attacks from the empirical studies [21,32] that demonstrate the current security practices fail to prevent such attacks as location hijacking, behavioral tracking, leaking cookies, and sniffing browsing history. Jang et al. [21] report on both explicit and implicit flows exploited in the empirical studies. Further, inlining by security-enhanced interpreting is a particularly suitable choice for tracking information flow in JavaScript, because alternative approaches to inlining suffer from scalability problems, as discussed in Section 5.

Our instantiation shows how to deploy *JSFlow* [19,18], an information-flow monitor for JavaScript by Hedin et al., via browser extension, via web proxy, and via suffix proxy (web service). We report on security and performance experiments that illustrate successful deployment of a JavaScript information-flow monitor with the different architectures.

2   Architectures

This section presents the architectures for inlining security monitors. We describe four different architectures and report on security considerations, pros and cons, including how the architectures reflect the demands of the different stakeholders. In the following we contrast the needs of the private user and the corporate user; the latter representing the network and system administrators as well.

2.1   Browser Extension

Modern browsers allow for the functionality of the browser to be enriched via *extensions*. By deploying the security monitor via a browser extension it is possible to enforce properties not normally enforced by browsers. A browser extension is a program that is installed into the browser in order to change or enrich the functionality. By employing a method pioneered by Zaphod [31] it is possible to use the monitor as JavaScript engine. The basic idea is to turn off JavaScript and have the extension traverse the page once loaded using the monitor to execute the scripts. This method leverages that the implementation language for extensions and the monitor is JavaScript.

It is illuminating to contrast deployment via browser extension with directly instrumenting the browser (e.g., [40,22,17]). While the latter may provide performance benefits it is also monitor specific. Each monitor leads to a different instrumented browser, which is a significant undertaking. In this sense, browser instrumentation is comparable to implementation of a new monitor. The browser extension, on the other hand, is parametric over the monitor allowing different monitors to be used with the same extension. While potentially slower than an instrumented browser it also offers greater flexibility.

Security considerations From a security perspective, one of the main benefits of this deployment method is strong security guarantees. Since the JavaScript engine is turned off, no code is executed unless explicitly done by the extension. During execution the scripts are passed as data to the monitor, and are only able to influence the execution environment implemented by the monitor and not the general execution environment. This ensures the integrity of the monitor and complete mediation. In addition, this also guarantees that the deployment method is sound given that the monitor is sound.

However, by running the monitor as an extension, the monitor is run with the same privileges as the browser. Compared to the other methods of deployment this means that a faulty monitor not only jeopardizes the property enforced by the monitor, but might jeopardize the integrity of the entire browser.

Pros and cons Regardless of whether the user is private or corporate, browser extensions provide a simple install-once deployment method. From the corporate perspective, central management of the extension and its policies can easily be incorporated into standard system maintenance procedures. Important for the private user, the fact that the extension is installed locally in the browser of the user makes it possible to give the user direct control over what security policies to enforce on the browsed pages without relying on and trusting other parties.

A general limitation of this approach is that browser extensions are browser specific. This is less of an issue for corporate users than for private users. In the former case it is common that browser restrictions are already in place, and corporations have the assets to make sure that extensions are available for the used platform. In the latter case, a private user may be discouraged by restrictions imposed by the extension.

2.2 Web Proxy

Deployment via browser extension entails being browser dependent and running the security monitor with elevated privileges. The web proxy approach addresses these concerns by including the monitor in the page, modifying any scripts on the page to ensure they are run by the monitor. All modern browsers support relaying all requests through a proxy. A proxy specific to relaying HTTP requests is referred to as a web proxy. The web proxy acts as a man-in-the-middle, making requests on behalf of the client. In the process, the proxy can modify both the request and the response, making it a convenient way to rewrite the response to include the monitor in each page. Doing so makes the method more intrusive to the HTML content, but less intrusive to the browser.

Security considerations For the monitor to guarantee security, all scripts bundled with the page must be executed by the monitor. The scripts can either be inline, i.e., included as part of the HTML page, or external, i.e., referenced in the HTML page to be downloaded from an external source. Inline scripts appear both in the form of script-tags as well as inline event handlers, e.g., onclick or onload. Apart from including the monitor in all browsed pages, all scripts, whether inline or external, must be rewritten by the web proxy to be executed by the monitor.

External scripts are rewritten in their entirety, whereas inline scripts must be identified within the page and rewritten them individually. As opposed to a browser extension that replaces the JavaScript engine, the monitor is executed by the engine of the browser in the context of the page. This is the same context in which all scripts bundled with the page are normally executed.

Unlike deployment via extension, omissions in this process breaks complete mediation, which risks undermining the integrity of the monitor; any script not subjected to the rewriting process is run in the same execution environment as the monitor. This assumes that there are no exploitable differences between the HTML parser used for rewriting and the parser of the browser. While this might be an issue in current browsers, with the introduction of standardized parsing in HTML5 we believe that this is a transient problem.

Under the assumption that all scripts are rewritten appropriately complete mediation is achieved. Complete mediation is required for both integrity and soundness, while the two latter are strongly related. Soundness is guaranteed by the soundness of the underlying monitor and complete mediation, given that the integrity of the monitor is guaranteed. This must, however, be the case, since the soundness of the monitor guarantees that no scripts executed by the monitor are able to jeopardize the integrity of the monitor. Thus, threats against the integrity of the monitor must come from scripts not run by the monitor, contradicting the assumption of complete mediation.

Unlike deployment via extension, special consideration is required for HTTPS connections, as HTTPS is designed to prevent the connection from being eavesdropped or modified in transit. To solve this the web proxy must establish two separate HTTPS connections, one with the client and one with the target. The client's request is passed on to the connection with the target and the rewritten response to the client. This puts considerable trust in the proxy, since the proxy has accesses to all information going to and from the user, including potentially sensitive or secret data. In addition, access to the unencrypted data significantly simplifies tampering unless additional measures are deployed. Whether including the proxy into the trusted computing base is acceptable or not depends on the situation.

Pros and cons In the corporate setting deployment via web proxy is appealing; it is common to use corporate proxies for filtering, which means that the infrastructure is already in place and trusted. Additionally, the use of proxies allows for easy central administration of security policies.

For the private user, however, the situation is different. Even though important considerations of extensions are addressed, e.g., browser dependency, and monitor privilege increase, adding the proxy to the trusted computing base might be a significant issue. Unless the private user runs and administers the proxy himself he might have little reason to trust the proxy with the ability to access all communicated information. This is especially true when the user visits web sites that he trusts more than the proxy. In such cases it could make sense to turn off the proxy, which, while possible, requires reconfiguring the browser.

2.3 Suffix Proxy (Service)

The extension and the web proxy deployment methods unconditionally applies the monitor to all visited pages. Suffix proxies can be used to provide selective monitoring, i.e., where the user can select when to use the monitor. Suffix proxies can be thought of as a service that allows the user to select which pages to proxy on demand — only pages visited using the suffix proxy will be subjected to proxying.

A *suffix proxy* is a specialized web proxy, with a different approach to relaying the request. The suffix proxy takes advantage of the *domain name system (DNS)* to redirect the request to it. Wildcard domain names allow all requests to any subdomain of the domain name to resolve to a single domain name, i.e., in DNS terms *.proxy.domain ⇒ proxy.domain*.

Typically, the user navigates to a web application associated with the proxy and enters the target URL, e.g., `http://google.com/search?q=sunrise`, in an input field. To redirect the request to the proxy, the target domain name is altered by appending the domain name of the proxy, making the target domain a subdomain of the proxy do-main, e.g., `http://google.com.proxy.domain/search?q=sunrise`. The suffix proxy is set up so that all requests to any subdomain are directed to the proxy domain. A web application on the proxy domain is set up to listen for such subdomain requests. When a request for a subdomain is registered, it is intercepted by the web appli-cation. The web application strips the proxy domain from the URL, leaving the original target URL, and makes the request on behalf of the client. As with the web proxy, re-laying the request to the target URL gives the suffix proxy an opportunity to modify and include the monitor in the response.

Security considerations In the suffix proxy, not only the content is rewritten but also the headers of the incoming request and the returned response. Certain headers, like the *Host* and *Referrer* headers of the request, include the modified domain name and need to be rewritten to make the proxy transparent. Similarly, in the response, some headers, for instance *Location*, contain the unmodified target URL and need to be rewritten to include the monitor domain.

As the web proxy, the suffix proxy must ensure that all scripts bundled with a page are executed by the monitor. The procedure to rewrite scripts is much the same as for the web proxy. However, in order to guarantee complete mediation, the suffix proxy must also rewrite the URLs to external scripts to include the proxy domain, in addition to rewriting inline scripts in a page. Otherwise the script will not be requested through the monitor which will prevent it from being rewritten and thereby it will execute along-side the monitor.

Identical to the web proxy, soundness and integrity is guaranteed by complete me-diation together with soundness of the monitor. See Section 2.2 above for a longer discussion.

A consequence of modifying the domain name is that the domain of the target URL no longer matches the modified URL, making them two separate origins as per the same-origin policy. This implies that all information in the browser specific to the target origin, e.g., cookies and local storage, are no longer associated with the modified origin, and vice versa. This results in a clean separation between the proxied and unproxied content.

Altering the domain name has another interesting effect on the same-origin policy. Modern web browsers allow relaxing of the same-origin policy for subdomains. Docu-ments from different subdomains of the same domain can relax their domains by setting the `document.domain` attribute to their common domain. In doing so, they set aside the restrictions of the same-origin policy and can freely access each others resources across subdomains. This means that two pages of separate origins loaded via the proxy, each relaxing their domain attribute to the domain of the proxy, can access each others resources across domains. This is problematic for monitors that rely on the same-origin policy to enforce separation between origins. However, the flexibility of disabling the same-origin policy opens up for monitors aimed at replacing the same-origin policy with policies that are more appropriate for the given scenario. For example, given es-tablished trust between a number of sites it is possible for a monitor to disable the same-origin policy between these sites, while leaving it enabled, or even strengthened (via, e.g., information flow tracking), for any sites outside the set of mutually trusting sites. Note that a monitor is always able to refuse relaxation by preventing scripts from changing the domain attribute; what the suffix proxy enables is the ability for monitors to modify and even disable the same-origin policy by allowing JavaScript to relax the domain. Clearly, if this is done, care must be taken not to introduce any security breaches.

Similar to the web proxy, HTTPS requires special consideration. For the suffix proxy, however, the situation is slightly simpler. Given that the suffix proxy builds on DNS wildcards, it is sufficient to issue a certificate for all subdomains of the proxy domain, e.g., `*.proxy.domain`. Such a wildcard certificate is valid for all target URLs relayed through the proxy.

Pros and cons In the corporate setting the suffix proxy does not offer any advantages over a standard web proxy. Giving corporate users control over the decision whether or not to use the proxy service opens up for mistakes.

From the perspective of a private user suffix proxies can be very appealing. Given that the suffix proxy is hosted by a trusted party, e.g., the user's ISP, the proxy can provide additional security for any web page. At the same time the user retains simple control over which pages are proxied. At any point, the user can opt out from the proxy service by not using it.

On the technical side, while sharing a common foundation, there are several differences between a suffix proxy compared to a traditional web proxy. The differences lie, not in how the monitor is included in the page, but in the way the proxy is addressed. A consequence of the use of wildcard domain names is that the suffix proxy requires somewhat more rewriting than the web proxy in order to capture all requests.

Additionally, the suffix proxy can ensure that only resources relevant to security are relayed via the proxy, whereas a traditional web proxy must cover all requests. This both reduces the load on the proxy service, as well as the overhead for the end user, thus benefiting both the user and the service provider. This is not possible in a web proxy, that must relay all requests, but a suffix proxy provides the means to do so.

2.4 Integrator

Modern web pages make extensive use of third-party code to add features and functionality to the page. The code is retrieved from external resources in the form of JavaScript libraries. The third-party code is considered to be part of the document and is executed in the same context as any other script included in the document. Executing the code in the context of the page gives the code full access to all the information of the page, including sensitive information such as form data and cookies. Granting such access requires that the code integrator must trust the library not to abuse this privilege. To a developer, an appealing alternative is to run untrusted code in the monitored context, while running trusted code outside of the monitor.

Integrator-driven monitor inclusion is suitable for web pages that make use of third-party code. The security of the information contained on the web page relies not only on the web page itself, but also on the security of all included libraries. To protect against malicious or compromised libraries, an integrator can execute part of, or all of the code in the monitor. Unlike the other deployment alternatives, that consider all code as untrusted, this approach requires a line to be drawn between trusted and untrusted code. The code executing outside of the monitor is trusted with full access to the page and the monitor state, and the untrusted code will be executed by the monitor, restricted from accessing either. This can be achieved by manually including the monitor in the page and loading the third-party code either through the suffix proxy, or from cached rewritten versions of the code. This approach allows for a well defined, site-specific policy specification. The monitor is set up and configured with policies best suiting the need of the site.

An important aspect of integrator-driven monitor inclusion is the interaction between trusted and untrusted code. The trusted code executing outside the monitor can interact with the code executed in the monitor. This way, the trusted code can share specific information with the library, that the library requires to execute. There are different means of introducing this information to the monitor. The most rudimentary solution is to evaluate expressions in the monitor, containing the information in a serialized form. The monitor can also provide an API for reading and writing variables, or calling functions in the monitor. This simplifies the process and makes it less error-prone. A more advanced solution is a set of shared variables that are bidirectionally reflected from one context to the other when their values are updated.

Security considerations One security consideration that arises is the implication of sharing information between the trusted and untrusted code. It might be appealing to simplify sharing of information between the two by reflecting a set of shared variables of one into the other. However, automatically reflecting information from one context to the other, will have severe security implications in terms of confidentiality as well as monitor integrity. If the execution of trusted code depends on a shared variable, the untrusted code can manipulate the value to control the execution. Thus, for security reasons, any sharing of information with the untrusted code must be done manually by selectively and carefully introducing the information in the monitored context.

It should be noted that since the trusted code is running along side the monitor, it can access and manipulate the state of the monitor and thereby the state of the untrusted code. It is impossible for the monitored code to protect against such manipulation. The integrator approach allows web developers to make use of untrusted code in trusted pages in a secure way. Thus, regardless of manually or automatically wrapping the untrusted code it is the responsibility of the integrator to ensure complete mediation. In the former case by making sure to wrap all untrusted code, and in the latter case by ensuring that eventual parser differences do not compromise mediation. Since the integrator approach provides complete mediation for untrusted code only, it is important that the trusted code does not break the integrity of the monitor. If this would occur soundness cannot be guaranteed, since the integrity breach could potentially allow for the untrusted code to break out of the monitor. However, if the trusted code does not break the integrity of the monitor, the integrator approach guarantees soundness and integrity with respect to the untrusted code in a similar manner to the proxy approach, see Section 2.2.

Pros and cons This developer-centric approach gives the integrator full control over the configuration of the monitor and the policies to enforce. From the perspective of a user this approach is not intrusive to the browser, requires no setup or configuration, and provides additional security for the user's sensitive information. However, it also limits the user's control over which policies are applied to user information.

A benefit of the integrator-driven approach over the proxies is potential performance gains. While the proxies for all code on the page to run monitored, the integrator-driven approach lets the integrator select what is monitored and what is not.

As previously stated, sharing information between trusted and untrusted code in a secure manner requires manual interaction. This implies that the developer must to some degree understand the inner workings of the monitor and the implications of interacting with the monitor.

2.5   Summary of Architectures

We have discussed four architectures for deployment. The differences between the architectures decide which architecture is better suited for different stakeholders and situations. The first three architectures were targeted to end users and were distinguished by their appeal to corporate and private users, whereas the last architecture was targeted on code developers.

Deployment via extension offers potentially stronger security guarantees at the price of running the monitor with the privilege of the browser, while the web proxy and suffix proxy approached were more susceptible to mistakes in the rewriting procedure. In the extension failing to identify a script leads to the script not executing, while in the proxies unidentified scripts would execute alongside the monitor, potentially jeopardizing its integrity. However, deployment via proxies requires someone to run and administer the proxies. For the corporate user the corporation is a natural host for such services, while the private user might lack such a trusted 3rd party. See Table 1 for a summary of how the architectures best suit each stakeholder. The results of the table are not firm. Rather they are recommendations based on the properties of the deployment methods and the needs of the different stakeholders in general.

For the corporate user we argue that deployment via web proxy may be the most natural method: it allows for simple centralized administration and, since it is common to use corporate proxies, the infrastructure might already be in place. As a runner up, deployment via extension is a good alternative deployment method, while the suffix proxy is the least attractive solution from a corporate perspective. The latter allows the user to select when to use the service, which opens up for security issues in case the user forgets to use the service.

For the private user we argue that deployment via extension is the most appealing method: after an initial installation it allows for local administration without the need to run additional services or rely on trusted 3rd parties. An interesting alternative for the private user is to use the suffix proxy. For web sites the private user trusts less than the provider of the suffix proxy service, the suffix proxy allows for increased security on a per web site basis. The web proxy is the least attractive means of deployment for the private user. From the private user's perspective the web proxy offers essentially the same guarantees as the extension, while either encumbering her to run her own proxy or rely on a 3rd party.

Finally, for the developer using the monitor as a library provides the possibility to include untrusted code safely using the monitor, while allowing trusted code to run normally. This allows for security, while lowering the performance impact by only monitoring potentially malicious parts of the program.

Table 1. Suitability of architectures with respect to different stakeholders

	Browser extension	Web proxy	Suffix proxy	Integrator driven
Corporate user		✓		
Private user	✓		✓	
Developer				✓

3 Implementation

This section details our implementations of the architectures from Section 2. The code is readily available and can be obtained from the authors upon request.

3.1 Browser Extension

The browser extension is a Firefox extension based on Zaphod [31], a Firefox extension that allows for the use of experimental Narcissus [13] engine as JavaScript engine. When loaded, the extension turns off the standard JavaScript engine by disallowing JavaScript and listens for the DOMContentLoaded event. DOMContentLoaded is fired as soon as the DOM tree construction is finished. On this event the DOM tree is traversed twice. The first traversal checks every node for event handlers, e.g., onclick, and registers the monitor to handle them. The second traversal looks for JavaScript script nodes. Each found script node is pushed onto an execution list, which is then processed in order. For each script on the execution list, the source is downloaded and the monitor is used to execute the script; any dynamically added scripts are injected into the appropriate place on the execution list.

The downside of this way of implementing the extension is that the order in which scripts are executed is important. When web pages are loaded, the scripts of the pages are executed as they are encountered while parsing the web page. This means that the DOM tree of the page might not have been fully constructed when the scripts execute. Differences in the state of the DOM tree can be detected by scripts at execution time. Hence, to guarantee transparency the execution of scripts must occur at the same times in the DOM tree construction as they would have in the unmodified browser. This can be achieved using DOM MutationEvent [41] instead of the DOMContentLoaded event. The idea is to listen to any addition of script nodes to the DOM tree under the construction, and execute the script on addition. However, due to performance reasons the DOM MutationEvents are deprecated, and are being replaced with DOM MutationObserver [42]. It is unclear whether the MutationObserver can be used to provide transparency, since events are grouped together, i.e., the mutation observer will not necessarily get an event each time a script is added — to improve performance single events may bundle several modifications together.

However, the exact order of loading is not standardized and differs between browsers. This forces scripts to be independent of such differences. Thus, using the method of executing scripts on the `DOMContentLoaded` event is not necessarily a problem in practice.

Further, since extension run with the same privileges as the browser certain protection mechanism are in place to protect the browser from misbehaving extensions. Those restrictions may potentially clash with the selected monitor. One example of this is `document.write`. The effect of `document.write` is [23] to write a string into the current position of the document. For security reasons, extensions are prohibited from calling `document.write`. Intuitively, `document.write` writes into the character stream that is fed to the HTML parser, which can have drastic effects on the parsing of the page. In a monitor it is natural to implement `document.write` by at some point calling the `document.write` of the browser. The alternative is to fully implement `document.write`, which would entail taking the interaction between the content written by `document.write`, the already parsed parts of the page and the remaining page into account. The inability to provide full functionality of `document.write` does not jeopardize the security, as argued in the introduction. Rather, it may prevent certain pages to execute properly. The consensus in the community is that `document.write` has few valid use cases, all pertaining to the inclusion of various entities during page load (calling `document.write` on a fully loaded page overwrites the entire page). One arguably reasonable use of `document.write` to include style sheets that only work when JavaScript is enabled. Another common use of `document.write`, that is broadly considered bad style, is to include scripts *synchronously* onto the page. Both approaches work by executing `document.write` with very specific strings as parameters, e.g.,

```
document.write(
    '<script src="http://somesite.com/script.js"></' + 'script>')
```

In such cases it is a simple matter to identify the attempt at inclusion, and mimic the appropriate behavior.

The extension consists of 1200 lines of JavaScript and XUL code.

3.2   Web Proxy

The web proxy is implemented as an HTTP-server. When the proxy receives a request it extracts the target URL and in turn requests the content from the target. Before the response is delivered to the client, the content is rewritten to ensure that all JavaScript is executed by the monitor.

In HTML, where JavaScript is embedded in the code, the web proxy must first identify the inline code in order to rewrite it. Identifying inline JavaScript in HTML files is a complex task. Simple search and replace is not satisfactory due to the browser's error tolerant parsing of HTML-code, meaning that the browser will make a best-effort attempt to make sense of malformed fragments of HTML. It would require the search algorithm to account for all parser quirks in regard to malformed HTML, a task which is at least as complex as actually parsing the document. Consider the example below:

```
<script>0</script/> HTML </script>
<script>0</script./> alert(JavaScript) </script>
<p>a<sCript/"=/ src=//t.co/abcde a= >b</p></script c<p>d
```

The first line will be interpreted as a script followed by the text *HTML*, the second line as a script that alerts the string *JavaScript*, and the third will display *ac* and *d* in two separate paragraphs and load a script from an external domain.

In the web proxy, Mozilla's JavaScript-based HTML-parser *dom.js* [16], is used to parse the page. HTML parsing is standardized in the HTML5 specification. The dom.js parser is HTML5 compliant and parses the HTML the same as any HTML5 compliant browser. In the case that the browser is not HTML5 compliant, or if there are implementation flaws, there may be ways to circumvent the rewriting in the proxy by exploiting such parsing mismatches. However, as browser vendors are implementing according to the specification to an increasing degree, these types of attacks are to be less likely.

After the page has been parsed, the DOM-tree can be traversed to properly localize all inline script code. All occurrences of JavaScript code are rewritten as outlined in the code snipped below, wrapped in a call to the monitor. Because all instances of the modified script code will reference the monitor, the monitor must be added as the first script to be executed.

Rewriting JavaScript requires converting the source code to a string that can be fed to the monitor. The method JSON.stringify() provides this functionality and will properly escape the string to ensure that it is semantically equivalent when interpreted by the monitor. The code string is then enclosed in a call to the monitors interpreter, as shown below:

```
code = 'Monitor.eval(' + JSON.stringify(code) + ')';
```

The implementation consists of 256 lines of JavaScript code.

3.3 Suffix Proxy (Service)

The web proxy serves as a foundation for the suffix proxy. The suffix proxy adds with an additional step of rewriting to deal with external resources. Since the suffix proxy is referenced by altering the domain name of the target, the proxy must ensure that relevant resources, e.g., scripts, associated with the target page are also retrieved through the proxy. Resources with relative URLs requires no processing, as they are relative to the proxy domain and will by definition be loaded through the monitor. However, the URLs of resources targeting external domains must be rewritten to include the proxy domain. Similarly, links to external pages must include the domain of the proxy for the monitor. The external references are identified in the same manner as inline JavaScript, by parsing the HTML to a DOM-tree and traversing the tree. When found, the URL is substituted using a regular expression.

Another difference to the web proxy relates to the use of non-standard ports. The web proxy will receive all requests regardless of the target port. The suffix proxy, on the other hand, only listens to the standard ports for HTTP and HTTPS, port 80 and 443 respectively. The port in a URL is specified in conjunction to, but not included in the domain. Hence any URLs specifying non-standard ports would attempt to connect to closed ports on the proxy server. A solution to this problem is to include the port as part of the modified domain name. To prevent clashing with the target domain or the proxy domain, the port number is included between the two, e.g., *http://target.domain.com.8080.proxy.domain/*. This does not clash with the target domain because the top domain of the target domain cannot be numeric. Neither does it clash with the proxy domain because it is still a subdomain of the proxy domain. The implementation consists of 276 lines of JavaScript code.

4 Instantiation

This section presents practical experiments made by instantiating the deployment architectures with the *JSFlow* [19,18] information-flow monitor. JSFlow is a tool that extends the formalization of a dynamic information flow tracker [20] to the full JavaScript language and its APIs. We briefly describe the monitor and discuss security and performance experiments.

Since the deployment approaches are parametric in the choice of monitor, we limit our interest to properties that relate to the approaches rather than the monitors. In particular, for the performance experiments we measure the time from issuing the request until the response is fully received, since the performance after that point depends entirely on the monitor, and for the security experiments we focus on results that are more generally interesting and not depending on the specific choice of monitor.

Monitor JSFlow is a dynamic information-flow monitor that tags values with runtime security labels. The security labels default to the origin of the data, e.g., user input is tagged *user*, but the labels can be controlled by the use of custom data attributes in the HTML. The default security policy is a strict version of the same-origin policy, where implicit flows, and flows via, e.g., image source attributes, are taken into account. Whenever a potential security violation has been encountered the monitor stops the execution with a security error. Implemented in JavaScript, the monitor supports full ECMA-262 (v5) [12] including the standard API and large parts of the browser-specific APIs such as the DOM APIs. JSFlow supports a wide variety of information-flow policies, including tracking of user input and preventing it from leaving the browser, as used in the security experiments below.

Security experiments Our experiments focus on password-strength checkers. After the user inputs a password, the strength of the password is computed according to some metric, and the result is displayed to the user, typically on a scale from *weak* to *strong*. This type of service is ubiquitous on the web, with service providers ranging from private web sites to web sites of national telecommunication authorities.

Clearly, the password needs to stay confidential; The strength of the password is irrelevant if the password is leaked. We have investigated a number of password-strength services. Our experiments identify services that enforce two types of policies: (i) allow the password to be sent back to the origin web site, but not to any other site (suitable for server-side checkers); and (ii) disallow the password to leave the browser (suitable for client-side checkers). The first type places trust on the service provider not to abuse the password, while the second type does not require such trust, in line with the *principle of least privilege* [39]. Note that these policies are indistinguishable from SOP's point of view because it is not powerful enough to express the second type.

One seemingly reasonable way to enforce the second type of policy is to isolate the service, i.e., prevent it from performing any communication. While effective, such a stern approach risks breaking the functionality of the service. It is common that pages employ usage statistics tracking such as Google Analytics. Google Analytics requires that usage information is allowed to be gathered and sent to Google for aggregation. Using information-flow tracking, we can allow communication to Google Analytics but with the guarantee that the password will not be leaked to it.

We have investigated a selection of sites[2] that fall into the first category, and a selection of sites[3] that fall into the second category. Of these, it is worth commenting on two sites, one from each category. Interestingly, the first site, `https://testalosenord.pts.se/`, is provided by the Swedish Post and Telecom Authority. The site contains a count of how many passwords have been submitted to the service, with over 1,000,000 tried passwords so far. We are in contact with the authority to help improve the security and usability of the service. The second, `http://www.getsecurepassword.com/CheckPassword.aspx`, is an example of a web site that uses Google Analytics. The monitor rightfully allows communication to Google while ensuring the password cannot be leaked anywhere outside the browser.

The benefit of the architectures for the scenario of password-strength checking is that users can get strong security guarantees either by installing an extension, using a web proxy, or a suffix proxy. In the latter two cases, the system and network administrators have a stake in deciding what policies to enforce. Further, the integrator architecture is an excellent fit for including a third-party password-strength checker into web pages of a service, say a social web site, with no information leaked to the third party.

Performance experiments Since the approaches are parametric in the choice of monitor, we are interested in evaluating the performance of each approach rather than the performance of the employed monitor. We measure the time from issuing the request until the response is fully received, as the performance after that point depends entirely on the monitor. We measure the average overhead introduced by architectures compared to a reference sample of unmodified requests. The overhead is measured against two of the password-strength checkers listed previously, namely `passwordmeter.com`, over HTTP, and `testalosenord.pts.se`, over HTTPS. This measures the additional overhead introduced by the deployment method. Three out of the four architectures are evaluated; the browser extension, the web proxy, and the suffix proxy. The overhead of the integrator architecture is specific to the page that implements it, and is therefore not comparable to the other three. The browser extension does not begin executing until the browser has received the page and has begun parsing it, therefore its response time is the same as with the extension disabled. Due to the rewriting mechanism being closely related, the web proxy and the suffix proxy show similar results.

---

[2] `https://testalosenord.pts.se/`, `http://www.lbw-soft.de/`, `http://www.inutile.ens.fr/estatis/password-security-checker/`, `https://passfault.appspot.com/password_strength.html`, and `http://geodsoft.com/cgi-bin/pwcheck.pl`.

[3] `http://www.getsecurepassword.com/CheckPassword.aspx`, `http://www.passwordmeter.com/`, `http://howsecureismypassword.net`, `https://www.microsoft.com/en-gb/security/pc-security/password-checker.aspx`, `https://www.grc.com/haystack.htm`, and `https://www.my1login.com/password-strength-meter.php`.

Table 2. Architecture overhead passwordmeter.com

	Measurements (ms)	Average (ms)	Delta (ms)	Overhead (%)
Reference	434, 420, 445, 443	435	0	0%
Browser extension	434, 420, 445, 443	435	0	0%
Web proxy	638, 690, 681, 781	697	+262	+60.2%
Suffix proxy	663, 775, 689, 694	705	+270	+62.0%

Table 3. Architecture overhead testalosenord.pts.se

Proxy	Measurements (ms)	Average (ms)	Delta (ms)	Overhead (%)
Reference	114, 240, 103, 104	140	0	0%
Browser extension	114, 240, 103, 104	140	0	0%
Web proxy	372, 308, 311, 305	324	+184	+131.4%
Suffix proxy	316, 314, 324, 333	321	+181	+129.2%

The tests were performed in the Firefox browser on the Windows 7 64 bit SP1 operating system on a machine with an Intel Core i7-3250M 2.9 GHz CPU, and 8 GB of memory.

5   Related Work

We first discuss the original work on reference monitors and their inlining, then inlining for information flow, and, finally, inlining security checks in the context of JavaScript.

Inlined reference monitors Anderson [3] introduces reference monitors and outlines the basic principles, recounted in Section 1. Erlingsson and Schneider [15,14] instigate the area of inlining reference monitors. This work studies both enforcement mechanisms and the policies that they are capable of enforcing, with the focus on safety properties. Inlined reference monitors have been proposed in a variety of languages and settings: from assembly code [15] to Java [10,11,9].

Ligatti et al. [25] present a framework for enforcing security policies by monitoring and modifying programs at runtime. They introduce *edit automata* that enable monitors to stop, suppress, and modify the behavior of programs.

Inlining for secure information flow Language-based information-flow security [37] features work on inlining for secure information flow. Secure information flow is not a safety property [29], but can be approximated by safety properties (e.g., [6,38,4]).

Chudnov and Naumann [7] have investigated an inlining approach to monitoring information flow in a simple imperative language. They inline a flow-sensitive hybrid monitor by Russo and Sabelfeld [36]. The soundness of the inlined monitor is ensured by bisimulation of the inlined monitor and the original monitor.

Magazinius et al. [28] cope with dynamic code evaluation instructions by inlining on-the-fly. Dynamic code evaluation instructions are rewritten to make use of auxiliary functions that, when invoked at runtime, inject security checks into the available string.

The inlined code manipulates shadow variables to keep track of the security labels of the program's variables. In similar vein, Bello and Bonelli [5] investigate on-the-fly inlining for a dynamic dependency analysis. However, there are fundamental limits in the scalability of the shadow-variable approach. The execution of a vast majority of the JavaScript operations (with the prime example being the + operation) is dependent on the types of their parameters. This might lead to coercions of the parameters that, in turn, may invoke such operations as toString and valueOf. In order to take any side effects of these methods into account, any operation that may case coercions must be wrapped. The end result of this is that the inlined code ends up emulating the interpreter, leaving no advantages to the shadow-variable approach.

Inlining for secure JavaScript Inlining has been explored for JavaScript, although focusing on simple properties or preventing against fixed classes of vulnerabilities. A prominent example in the context of the web is BrowserShield [34] by Reis et al. to instrument scripts with checks for known vulnerabilities.

Yu et al. [44] and Kikuchi et al. [24] present an instrumentation approach for JavaScript in the browser. Their framework allows instrumented code to encode edit automata-based policies.

Phung et al. [33] and Magazinius et al. [27] develop secure wrapping for self-protecting JavaScript. This approach is based on wrapping built-in JavaScript methods with secure wrappers that regulate access to the original built-ins.

Agten et al. [2] present JSand, a server-driven client-side sandboxing framework. The framework mediates attempts of untrusted code to access resources in the browser. In contrast to its predecessors such as ConScript [30], WebJail [1], and Contego [26], the sandboxing is done purely at JavaScript level, requiring no browser modification.

Despite the above progress on inlining security checks in JavaScript, achieving information-flow security for client-side JavaScript by inlining has been out of reach for the current methods [40,8,43,21,17] that either modify the browser or perform the analysis out-of-the-browser.

6 Conclusions

Different stakeholders have different interests in the security of web applications. We have presented architectures for inlining security monitors, to take into account the security goals of the users, system and network administrators, and service providers and integrators. We achieve great flexibility in the deployment options by considering security monitors implemented as security-enhanced JavaScript interpreters. The architectures allow deploying such a monitor in a browser extension, web proxy, or web service. We have reported on the security considerations and on the relative pros and cons for each architecture. We have applied the architectures to inline an information-flow security monitor for JavaScript. The security experiments show the flexibility in supporting the different policies on the sensitive information from the user. The performance experiments show reasonable overhead imposed by the architectures.

Future work is focused on three promising directions. First, JavaScript may occur outside script elements, e.g., as part of css, SVG or Flash. Ignoring JavaScript outside script elements potentially opens up for bypassing the security policies. One possible solution to this is to disallow JavaScript from occurring outside normal script tags, either by removing it of turning off the enabling features (e.g., Flash). Even though such a method might seem drastic it is conceivable due to the limited proliferation of JavaScript outside normal scripts. Nevertheless, to provide a more complete solution, we aim to investigate extending the approaches presented in this paper to handle such JavaScript. Second, recall that the integrator architecture relies on the developer to establish communication between the monitored and unmonitored code. With the goal to relieve the integrator from manual efforts, we develop a framework for secure communication that provides explicit support for integrating and monitored and unmonitored code. Third, we pursue instantiating the architectures with a monitor for controlling network communication bandwidth.

Acknowledgments. Thanks are due to Christian Hammer for useful feedback. This work was funded by the European Community under the ProSecuToR and WebSand projects and the Swedish agencies SSF and VR.

References

1. Acker, S.V., Ryck, P.D., Desmet, L., Piessens, F., Joosen, W.: Webjail: least-privilege integration of third-party components in web mashups. In: Proc. of ACSAC 2011 (2011)
2. Agten, P., Acker, S.V., Brondsema, Y., Phung, P.H., Desmet, L., Piessens, F.: JSand: complete client-side sandboxing of third-party JavaScript without browser modifications. In: Zakon, R.H. (ed.) ACSAC 2012, pp. 1–10. ACM (2012)
3. Anderson, J.P.: Computer security technology planning study. Technical report, Deputy for Command and Management System, USA (1972)
4. Austin, T.H., Flanagan, C.: Efficient purely-dynamic information flow analysis. In: Proc. ACM Workshop on Programming Languages and Analysis for Security, PLAS (June 2009)
5. Bello, L., Bonelli, E.: On-the-fly inlining of dynamic dependency monitors for secure information flow. In: Barthe, G., Datta, A., Etalle, S. (eds.) FAST 2011. LNCS, vol. 7140, pp. 55–69. Springer, Heidelberg (2012)
6. Boudol, G.: Secure information flow as a safety property. In: Degano, P., Guttman, J., Martinelli, F. (eds.) FAST 2008. LNCS, vol. 5491, pp. 20–34. Springer, Heidelberg (2009)
7. Chudnov, A., Naumann, D.A.: Information flow monitor inlining. In: Proc. of CSF 2010 (2010)
8. Chugh, R., Meister, J.A., Jhala, R., Lerner, S.: Staged information flow for JavasCript. In: Hind, M., Diwan, A. (eds.) PLDI, pp. 50–62. ACM (2009)
9. Dam, M., Guernic, G.L., Lundblad, A.: Treedroid: a tree automaton based approach to enforcing data processing policies. In: Proc. of ACM CCS 2012, pp. 894–905 (2012)
10. Dam, M., Jacobs, B., Lundblad, A., Piessens, F.: Security monitor inlining for multithreaded java. In: Drossopoulou, S. (ed.) ECOOP 2009. LNCS, vol. 5653, pp. 546–569. Springer, Heidelberg (2009)
11. Dam, M., Jacobs, B., Lundblad, A., Piessens, F.: Provably correct inline monitoring for multithreaded java-like programs. Journal of Computer Security 18(1), 37–59 (2010)
12. ECMA International. ECMAScript Language Specification, Version 5 (2009)
13. B. Eich. Narcissus—JS implemented in JS (2011),
    http://mxr.mozilla.org/mozilla/source/js/narcissus/

14. Erlingsson, U.: The inlined reference monitor approach to security policy enforcement. PhD thesis, Cornell University, Ithaca, NY, USA (2004)
15. Erlingsson, U., Schneider, F.B.: Sasi enforcement of security policies: a retrospective. In: Proc. of NSPW 1999, pp. 87–95 (1999)
16. Gal, A.: dom.js, `https://github.com/andreasgal/dom.js`
17. Groef, W.D., Devriese, D., Nikiforakis, N., Piessens, F.: Flowfox: a web browser with flexible and precise information flow control. In: Proc. of ACM CCS 2012 (October 2012)
18. Hedin, D., Birgisson, A., Bello, L., Sabelfeld, A.: JSFlow. Software release (September 2013), Located at `http://chalmerslbs.bitbucket.org/jsflow`
19. Hedin, D., Birgisson, A., Bello, L., Sabelfeld, A.: JSFlow: Tracking Information Flow in JavaScript and its APIs. In: SAC. ACM (March 2014)
20. Hedin, D., Sabelfeld, A.: Information-flow security for a core of JavaScript. In: Proc. IEEE Computer Security Foundations Symposium, pp. 3–18 (June 2012)
21. Jang, D., Jhala, R., Lerner, S., Shacham, H.: An empirical study of privacy-violating information flows in JavaScript web applications. In: Proc. of ACM CCS 2010 (October 2010)
22. Just, S., Cleary, A., Shirley, B., Hammer, C.: Information Flow Analysis for JavaScript. In: Proc. of PLASTIC 2011 (2011)
23. Kesselman, J.: Document Object Model (DOM) Level 2 Core Specification (2000)
24. Kikuchi, H., Yu, D., Chander, A., Inamura, H., Serikov, I.: Javascript instrumentation in practice. In: Ramalingam, G. (ed.) APLAS 2008. LNCS, vol. 5356, pp. 326–341. Springer, Heidelberg (2008)
25. Ligatti, J., Bauer, L., Walker, D.: Edit automata: Enforcement mechanisms for run-time security policies. International Journal of Information Security 4, 2–16 (2005)
26. Luo, T., Du, W.: Contego: Capability-based access control for web browsers - (short paper). In: McCune, J.M., Balacheff, B., Perrig, A., Sadeghi, A.-R., Sasse, A., Beres, Y. (eds.) Trust 2011. LNCS, vol. 6740, pp. 231–238. Springer, Heidelberg (2011)
27. Magazinius, J., Phung, P.H., Sands, D.: Safe wrappers and sane policies for self protecting javascript. In: Aura, T., Järvinen, K., Nyberg, K. (eds.) NordSec 2010. LNCS, vol. 7127, pp. 239–255. Springer, Heidelberg (2012)
28. Magazinius, J., Russo, A., Sabelfeld, A.: On-the-fly inlining of dynamic security monitors. Computers & Security 31(7), 827–843 (2012)
29. McLean, J.: A general theory of composition for trace sets closed under selective interleaving functions. In: Proc. IEEE Symp. on Security and Privacy, pp. 79–93 (May 1994)
30. Meyerovich, L.A., Livshits, V.B.: Conscript: Specifying and enforcing fine-grained security policies for javascript in the browser. In: Proc. of IEEE S&P 2010 (2010)
31. Mozilla Labs. Zaphod add-on for the Firefox browser (2011), `http://mozillalabs.com/zaphod`
32. Nikiforakis, N., Invernizzi, L., Kapravelos, A., Van Acker, S., Joosen, W., Kruegel, C., Piessens, F., Vigna, G.: You are what you include: large-scale evaluation of remote JavaScript inclusions. In: Proc. of ACM CCS 2012, pp. 736–747 (October 2012)
33. Phung, P.H., Sands, D., Chudnov, A.: Lightweight self-protecting javascript. In: Proc. of ASIACCS 2009, pp. 47–60 (2009)
34. Reis, C., Dunagan, J., Wang, H.J., Dubrovsky, O., Esmeir, S.: Browsershield: Vulnerability-driven filtering of dynamic html. ACM Trans. Web 1(3), 11 (2007)
35. Rushby, J.M.: Design and verification of secure systems. In: Proc. SOSP 1981 (1981)
36. Russo, A., Sabelfeld, A.: Dynamic vs. static flow-sensitive security analysis. In: Proc. IEEE Computer Security Foundations Symposium, pp. 186–199 (July 2010)
37. Sabelfeld, A., Myers, A.C.: Language-based information-flow security. IEEE J. Selected Areas in Communications 21(1), 5–19 (2003)

38. Sabelfeld, A., Russo, A.: From dynamic to static and back: Riding the roller coaster of information-flow control research. In: Pnueli, A., Virbitskaite, I., Voronkov, A. (eds.) PSI 2009. LNCS, vol. 5947, pp. 352–365. Springer, Heidelberg (2010)
39. Saltzer, J.H., Schroeder, M.D.: The protection of information in computer systems. Proc. of the IEEE 63(9), 1278–1308 (1975)
40. Vogt, P., Nentwich, F., Jovanovic, N., Kirda, E., Kruegel, C., Vigna, G.: Cross-site scripting prevention with dynamic data tainting and static analysis. In: Proc. of NDSS (February 2007)
41. W3C. Document Object Model (DOM) Level 3 Events Specification, http://www.w3.org/TR/DOM-Level-3-Events/
42. W3C. DOM4 W3C Working Draft 6, http://www.w3.org/TR/dom/
43. Yip, A., Narula, N., Krohn, M., Morris, R.: Privacy-preserving browser-side scripting with BFlow. In: EuroSys 2009, pp. 233–246. ACM, New York (2009)
44. Yu, D., Chander, A., Islam, N., Serikov, I.: JavaScript instrumentation for browser security. In: Proc. ACM Symp. on Principles of Programming Languages, pp. 237–249. ACM (2007)

Automatic and Robust Client-Side Protection for Cookie-Based Sessions

Michele Bugliesi, Stefano Calzavara, Riccardo Focardi, and Wilayat Khan

Università Ca' Foscari Venezia
{michele,calzavara,focardi,khan}@dais.unive.it

Abstract. Session cookies constitute one of the main attack targets against client authentication on the Web. To counter that, modern web browsers implement native cookie protection mechanisms based on the Secure and HttpOnly flags. While there is a general understanding about the effectiveness of these defenses, no formal result has so far been proved about the security guarantees they convey. With the present paper we provide the first such result, with a mechanized proof of noninterference assessing the robustness of the Secure and HttpOnly cookie flags against both web and network attacks. We then develop CookiExt, a browser extension that provides client-side protection against session hijacking based on appropriate flagging of session cookies and automatic redirection over HTTPS for HTTP requests carrying such cookies. Our solution improves over existing client-side defenses by combining protection against both web and network attacks, while at the same time being designed so as to minimise its effects on the user's browsing experience.

1 Introduction

Providing access to online content-rich resources such as those available in modern web applications requires tracking a user's identity through multiple requests. That, in turn, leads naturally to introduce the concept of *web session* to gather different HTTP(S) requests under the same identity, and implement a stateful, authenticated communication paradigm.

State information in web sessions is typically encoded by means of *cookies*: a cookie is a small piece of data generated by the server, sent to the user's browser and stored therein, for the browser to attach it automatically to all HTTP(S) requests to the server which registered it. If the cookie contains an adequately long string of random data, the server can effectively identify the client and restore its state, thus implementing a web session across different requests.

Cookie-based sessions are exposed to serious security threats, as the inadvertent disclosure of a session cookie provides an attacker with full capabilities of impersonating the client identified by that cookie. Indeed, cookie theft constitutes one of the most prominent web security attacks and several approaches have been proposed in the past to prevent and/or mitigate it [17,16,9,15]. Interestingly, this problem is so serious that modern web browsers implement native protection mechanisms based on the Secure and HttpOnly flags to shield session

J. Jürjens, F. Piessens, and N. Bielova (Eds.): ESSoS 2014, LNCS 8364, pp. 161–178, 2014.

cookies from unintended access by scripts injected within HTML code, as well as by sniffers tapping the client-server link of an HTTP connection. While there is a general understanding that these flags constitute an effective defense, no formal result has so far been proved about the security guarantees they convey.

Contributions. With the present paper we provide the first such result assessing the robustness of the `Secure` and `HttpOnly` cookie flag mechanisms with respect to a precise and rigorous attacker model, which captures both web threats (based on, e.g., code injection) and network attacks. We state our result in terms of *reactive noninterference* [7], a popular and widely accepted definition of information security, which provides strong, full-rounded protection against any (direct or indirect) information flow occurring in the browser. To carry out our mechanized proof, we extend Featherweight Firefox [6,5], a core model of a web browser developed in the Coq proof assistant [1], and we rely on Coq's facilities for interactive theorem proving to establish our result.

The security guarantees provided by our mechanized proof apply only to sessions that draw on appropriately flagged cookies. Clearly, however, poorly engineered websites that do not comply with the required flagging still expose their users to serious risks of session hijacking. Our analysis of the Alexa-ranked top 1000 popular websites gives clear evidence that such risks are far from remote, as the `Secure` and `HttpOnly` flags appear as yet to be largely ignored by web developers. As a countermeasure, we propose CookiExt, a browser extension that provides client-side protection against the theft of session cookies, based on appropriate flagging of such cookies and automatic redirection over HTTPS for HTTP requests carrying them.

We discuss the design of our Google Chrome implementation of CookiExt, and report on the experiments we carried out to evaluate the effectiveness of our approach. As we discuss in the related work section, CookiExt improves over existing client-side defenses by combining protection against both web and network attacks, while at the same time being designed so as to minimise its effects on the user's browsing experience.

Structure of the paper. Section 2 provides background material. Section 3 describes our formal model and the main theoretical results. Section 4 focuses on the practical aspects of session cookie security and presents CookiExt. Section 5 compares the present paper to related work. Section 6 concludes[1].

2   Background

2.1   Session Cookies: Attacks and Defenses

Web Attacks. Web browsers store cookies in their local storage and implement a simple protection mechanism based on the so-called "same-origin policy", whereby cookies registered by a given domain are made accessible only

---

[1] CookiExt and Coq scripts available at https://github.com/wilstef/secookie

to scripts retrieved from that same domain. Unfortunately, as it is well-known, the same-origin policy may be circumvented by a widespread form of code injection attacks known as *cross-site scripting* (XSS). In these attacks, a script crafted by the attacker is injected in a page originating from a trusted web site and thus acquires the privileges of that site [10]. As a result, the injected script is granted access to the DOM of the page, in particular to the Javascript object `document.cookie` containing the session cookies, which it can leak to the attacker's website.

Network Attacks. A network attacker may be able to fully inspect all the unencrypted traffic exchanged between the browser and the server. Though adopting HTTPS connections to encrypt network traffic would provide an effective countermeasure against eavesdropping, protecting session cookies against improper disclosure is tricky. On the one hand, many websites are still only *partially* deployed over HTTPS, and require special attention. In fact, cookies registered by a given domain are by default attached to *all* the requests to that domain: consequently, unless appropriate protection is put in place, loading a page over HTTPS may still leak session cookies, whenever the page retrieves additional contents (e.g., images or scripts) over an HTTP connection to the same domain. On the other hand, even websites which are *completely* deployed over HTTPS are vulnerable to session cookie theft, whenever an attacker is able to inject HTTP links to them in unrelated web pages [13].

Protection Mechanisms. Web development frameworks provide two main mechanisms to secure session cookies, based on the `Secure` and `HttpOnly` flags. The `HttpOnly` flag blocks any access to a cookie attempted by JavaScript or any other non-HTTP API, thus making cookies available only upon transmissions of HTTP(S) requests and thwarting XSS attacks. The `Secure` flag, in turn, informs the browser that a cookie may only be included in requests sent over HTTPS connections, thus ensuring that the cookie is always encrypted when transmitted from the client to the server. Since `Secure` cookies will never be attached to requests performed over HTTP connections, they are protected against the security flaws discussed above.

2.2 Formal Browser Models

Web browsers can be formalized in terms of constrained labelled transition systems known as *reactive* systems [6]. Intuitively, a reactive system is an event-driven state machine which waits for an input, produces a sequence of outputs in response, and repeats the process indefinitely.

Definition 1 (Reactive System [7]). *We define a reactive system as a tuple* $(\mathcal{C}, \mathcal{P}, \mathcal{I}, \mathcal{O}, \longrightarrow)$*, where* $\mathcal{C}$ *and* $\mathcal{P}$ *are disjoint sets of consumer and producer states respectively,* $\mathcal{I}$ *and* $\mathcal{O}$ *are disjoint sets of input and output events respectively. The last component,* $\longrightarrow$*, is a labelled transition relation over the set of states* $\mathcal{S} \triangleq \mathcal{C} \cup \mathcal{P}$ *and the set of labels* $\mathcal{A} \triangleq \mathcal{I} \cup \mathcal{O}$*, defined by the following clauses:*

1. $C \in \mathcal{C}$ *and* $C \xrightarrow{\alpha} Q$ *imply* $\alpha \in \mathcal{I}$ *and* $Q \in \mathcal{P}$;
2. $P \in \mathcal{P}$, $Q \in \mathcal{S}$ *and* $P \xrightarrow{\alpha} Q$ *imply* $\alpha \in \mathcal{O}$;

3. $C \in \mathcal{C}$ and $i \in \mathcal{I}$ imply $\exists P \in \mathcal{P} : C \xrightarrow{i} P$;

4. $P \in \mathcal{P}$ implies $\exists o \in \mathcal{O}, Q \in \mathcal{S} : P \xrightarrow{o} Q$.

Defining a notion of information security for reactive systems requires one to identify how input events affect the output events generated in response. We define (possibly infinite) *streams* of events, coinductively, as the largest set generated by the following productions: $S := [] \mid s :: S$, where s ranges over stream elements. Then, we characterize the behaviour of a reactive system as a transformation of a given input stream into a corresponding output stream.

Definition 2 (Trace). *For an input stream I, a reactive system in a given state Q computes an output stream O iff the judgement $Q(I) \Rightarrow O$ can be derived by the following inference rules:*

$$\text{(C-Nil)} \over C([]) \Rightarrow []$$

$$\text{(C-In)} \qquad \frac{C \xrightarrow{i} P \qquad P(I) \Rightarrow O}{C(i :: I) \Rightarrow O}$$

$$\text{(C-Out)} \qquad \frac{P \xrightarrow{o} Q \qquad Q(I) \Rightarrow O}{P(I) \Rightarrow o :: O}$$

A reactive system generates the trace (I, O) *if we have $Q_0(I) \Rightarrow O$, where Q_0 is the initial state of the reactive system.*

2.3 Reactive Noninterference

Given the previous definition, we can introduce an effective notion of information security based on the theory of *noninterference*. We presuppose a pre-order of security labels $(\mathcal{L}, \sqsubseteq)$ and characterize the power of an observer in terms of the labels $l \in \mathcal{L}$, where higher labels correspond to higher power.

Definition 3 (Reactive Noninterference [7]). *A reactive system is noninterferent if for all labels l and all its traces (I, O) and (I', O') such that $I \approx_l I'$, one has $O \approx_l O'$.*

The notation $S \approx_l S'$ identifies a similarity relation on streams, which corresponds to the inability of an observer at level l to distinguish S from S'. As discussed in [7], different definitions of stream similarity correspond to different sensible notions of information security. We focus on a very natural definition, which gives rise to a practically useful (termination-insensitive) notion of noninterference called *indistinguishable security*.

Definition 4 (Stream Similarity). *We let $\approx_l$ be the largest relation closed under the following inference rules:*

$$\text{(S-Nil)} \over [] \approx_l []$$

$$\text{(S-Vis)} \qquad \frac{visible_l(s) \qquad visible_l(s') \qquad s \approx_l s' \qquad S \approx_l S'}{s :: S \approx_l s' :: S'}$$

$$\text{(S-InvisL)} \qquad \frac{\neg visible_l(s) \qquad S \approx_l S'}{s :: S \approx_l S'}$$

$$\text{(S-InvisR)} \qquad \frac{\neg visible_l(s') \qquad S \approx_l S'}{S \approx_l s' :: S'}$$

The definition is parametric with respect to a visibility and a similarity relations for individual stream elements. Different instantiations of these relations entail different security guarantees, as we discuss in Section 3.3.

3   Formalizing Session Security

We continue with an outline of our mechanized formal proof of reactive noninterference for properly flagged session cookies under the currently available browser protection mechanisms. To ease readability, we keep the presentation informal (though rigorous) whenever possible: full details can be found in the online Coq scripts at `https://github.com/wilstef/secookie`.

3.1   Extending Featherweight Firefox

Featherweight Firefox (FF) is a core model of a web browser, developed in the Coq proof assistant [6,5]. Despite its name, the model is not tailored specifically around Firefox. Instead, it provides a fairly rich subset of the main functionalities of any standard web browser, including multiple browser windows, cookies, HTTP requests and responses, basic HTML elements, a simple Document Object Model, and some of the essential features of JavaScript.

FF is a reactive system: input events can either originate from the user or from the network, and output events can similarly be sent to the user or to the network. In particular, the model defines how the browser reacts to each possible input by emitting (a sequence of) outputs in response.

We extend FF with a number of new features, to include (i) support for HTTPS communication; (ii) a more accurate management of the browser cookie store to capture the `Secure` and `HttpOnly` flags with their intended semantics, and (iii) HTTP(S) redirects, a feature included in related models [2,3] which has been shown to have a significant impact on browser security.

The implementation of the extended model arises as expected, though it requires several changes to the existing framework to get a working Coq program. We just remark that HTTPS communication is modelled symbolically, by extending the syntax of input and output events to make it possible to discriminate between plain and encrypted exchanges (see below).

3.2   Threat Model

As anticipated, we characterize attackers in terms of a pre-order on security labels, which we define as follows.

Definition 5 (Security Labels and Order). *Let $\mathcal{D}$ be a denumerable set of domain names, ranged over by d. We define the set of security labels $\mathcal{L}$, ranged over by l, as the smallest set generated by the following grammar:*

$$l := \bot \mid \top \mid \mathsf{net} \mid \mathsf{http}(d) \mid \mathsf{https}(d).$$

We define $\sqsubseteq$ as the least reflexive relation over $\mathcal{L}$, with $\top$ as a top element, $\bot$ as a bottom element, and closed under the following inference rules:

(O-NetL)   (O-NetR)   (O-Https)

$\overline{\mathsf{net} \sqsubseteq \mathsf{https}(d)}$ $\overline{\mathsf{http}(d) \sqsubseteq \mathsf{net}}$ $\overline{\mathsf{http}(d) \sqsubseteq \mathsf{https}(d')}$

We can easily prove that $(\mathcal{L}, \sqsubseteq)$ is a pre-order, hence our definition is well-suited for the theory of reactive noninterference.

It is very natural to characterize the attacker power in terms of a security label in the previous pre-order. Specifically, level $\mathsf{http}(d)$ corresponds to a *web attacker* at d, which has no network capability and can only observe data sent to d itself. A *network attacker*, instead, resides at level net and is stronger than any web attacker, being able to inspect the contents of all the unencrypted network traffic. We additionally assume that a net-level attacker is able to observe the *presence* of any HTTPS request sent over the network, even though he does not have access to its contents. Finally, level $\mathsf{https}(d)$ corresponds to an even more powerful attacker, which has fully compromised the web server at d: this attacker has all the capabilities of a network attacker and can also decrypt all the encrypted traffic sent to d.

We anticipate that, by quantifying over all the possible inputs, our model will implicitly provide the attacker (at any level) with the ability to inject malicious contents on all websites, thus naturally capturing XSS attacks.

3.3   Noninterference for Session Cookies

The two security properties we target may informally be described as follows:

(1) the value of an `HttpOnly` cookie registered by a domain d can only be disclosed by an attacker at level $\mathsf{http}(d)$ or higher;
(2) the value of an `HttpOnly` and `Secure` cookie registered by a domain d can only be disclosed by an attacker at level $\mathsf{https}(d)$ or higher.

In both cases we target strong confidentiality guarantees, to ensure that the secrecy of a session cookie is protected against both explicit and implicit flows of information. Interestingly, we can uniformly characterize properties (1) and (2) in terms of reactive noninterference and carry out a single security proof which entails both. Specifically, we will show that a browser reacting to two input streams that are indistinguishable up to the values of the `HttpOnly` (and `Secure`, when conveyed over HTTPS) cookies attached to the streams' components, will produce indistinguishable output streams for any observer/attacker that is not the intended owner of these cookies.

Overview. Before delving into the technical details, we first provide an intuition about how existing attacks are captured by reactive noninterference. Consider a web attacker running the website `attacker.com` and take the following script snippet:

```
val = get_ck_val(document.cookie,"PHPSESSID");
for (x in val) {
   <contact http://attacker.com/leak?pos=x&char=val[x]>
}
```

The script retrieves the `document.cookie` object containing all the cookies accessible by the page to read the value of the `PHPSESSID` cookie (storing a session identifier), and then leaks each character of the cookie value to the attacker's website. If the script is injected into a response from `honest.com`, for instance through XSS, the attacker will be able to hijack the user's session.

Now notice that, if the cookie `PHPSESSID` is marked as `HttpOnly`, the previous attack will not work. In particular, irrespective of the value stored in the cookie, the observable output available to the attacker will always be the same, i.e., nothing. This ensures that both the value of the cookie and its length (an *implicit* flow of information) are not disclosed to the attacker, and it is thus safe to deem two HTTP(S) responses from `honest.com` as *similar* whenever they are identical up to the choice of the `PHPSESSID` cookie value. If the `HttpOnly` flag is not applied to the cookie, we cannot treat the two responses as similar, since the attacker would be able to draw a difference between them based on the outputs observable at `attacker.com`, thus violating reactive noninterference.

The reasoning can be generalized to the `Secure` flag and network attackers, with the proviso that any `Secure` cookie must be marked also as `HttpOnly` to be actually protected: the script above already highlights this point. Indeed, if the cookie `PHPSESSID` is only marked as `Secure`, the previous script could still leak over HTTP all the characters composing the cookie value, thus allowing a network attacker to draw a difference between two responses identical up to the cookies they set.

Formalization. We start by introducing some notation. We let URLs be defined by the productions: $url := \mathsf{blank} \mid \mathsf{url}(protocol, domain, path)$, where $protocol \in \{\mathsf{http}, \mathsf{https}\}$, and $domain$ and $path$ are arbitrary strings. We let uwi range over window identifiers, i.e., natural numbers serving as an internal representation of browser windows; similarly, we let $ncid$ range over network connection identifiers, which are needed in the browser model to match responses with their corresponding requests. A network connection identifier is a record with two fields: url, which contains the URL endpoint of the connection, and $value$, a natural number which uniquely identifies the connection. We use the dot operator "." to perform the lookup of a record field.

Output events. Output events are defined by the following, mostly self-explanatory productions:

$$o := \mathsf{ui\_window\_opened} \mid \mathsf{ui\_window\_closed}(uwi)$$
$$\mid \mathsf{ui\_page\_loaded}(uwi, url, doc) \mid \mathsf{ui\_page\_updated}(uwi, doc)$$
$$\mid \mathsf{ui\_error}(msg) \mid \mathsf{net\_doc\_req}(ncid, req)$$
$$\mid \mathsf{net\_script\_req}(ncid, req) \mid \mathsf{net\_xhr\_req}(ncid, req).$$

We define *visibility* for output events by means of the following inference rules:

(VO-NET)
$$\frac{url\_label(ncid) \sqcap net \sqsubseteq l \quad * \in \{doc, script, xhr\}}{visible_l(net\_*\_req(ncid, req))}$$

(VO-TOP)
$$\overline{visible_\top(o)}$$

The partial function $url\_label(\cdot)$ maps network connection identifiers to security labels as follows:

$$url\_label(ncid) = \begin{cases} https(d) & \text{if } \exists p : ncid.url = url(https, d, p) \\ http(d) & \text{if } \exists p : ncid.url = url(http, d, p) \end{cases}$$

The definition is consistent with the previous characterization of the attacker on the label pre-order. In particular, notice that both the encrypted and the unencrypted network traffic is visible to any attacker l such that $l \sqsupseteq net$.

We then define *similarity* for output events through the following rules:

(SO-CRYPT)
$$\frac{\exists d : url\_label(ncid) = https(d) \not\sqsubseteq l \quad * \in \{doc, script, xhr\}}{net\_*\_req(ncid, req) \approx_l net\_*\_req(ncid, req')}$$

(SO-REFL)
$$\overline{o \approx_l o}$$

In words, we are assuming that the attacker is able to fully analyse any plain output event it has visibility of, while the contents of an encrypted request can only be inspected by a sufficiently strong attacker, who is able to decrypt the message. We assume a randomized encryption scheme, whereby encrypting the same request twice always produces two different ciphertexts[2]. Notice that similar ($\approx_l$) output events must be sent to the same URL, i.e., we assume that the attacker is able to observe the recipient of any visible network event.

Input events. The treatment for input events is similar, but subtler. Again, we start by introducing some notation. We let network responses (*resp*) be defined as records with four fields: *del_cookies*, which is a set of names of cookies which should be deleted by the browser; *set_cookies*, which is a set of cookies which should be stored in the browser; *redirect_url*, which is an (optional) URL needed for HTTP(S) redirects; and *file*, which is the body of the response. Cookies, in turn, are ranged over by c and defined as records with six fields: a *name*, a *value*, a *domain*, a *path*, and two boolean flags *secure* and *httponly*.
Input events are then defined by the following productions:

$$i := \text{ui\_load\_in\_window}(uwi, url) \mid \text{ui\_close\_window}(uwi)$$
$$\mid \text{ui\_input\_text}(uwi, field, msg) \mid \text{net\_doc\_resp}(ncid, resp)$$
$$\mid \text{net\_script\_resp}(ncid, resp) \mid \text{net\_xhr\_resp}(ncid, resp).$$

We presuppose a further condition to rule out input events built around Secure cookies registered over HTTP. This corresponds to assuming the following

---

[2] This is a sound assumption for HTTPS, since it relies on the usage of short-term symmetric keys and attaches different sequence numbers to different messages.

condition for the last three clauses of the input-event productions above: whenever there exists a cookie $c \in resp.set\_cookies$ such that $c.secure = \mathsf{true}$, then $ncid.url$ must be of the form $\mathsf{url}(\mathsf{https}, d, p)$ for some d and p. Any input that does not satisfy this condition is clearly ill-formed, as Secure cookies received in the clear cannot be protected at the client side. Ruling out ill-formed inputs provides then the formal counterpart of having the browser simply reject them, declining the request to store the cookie. Indeed, while current browsers do not seem to implement this check, that is enforced by our browser extension (cf. Section 4).

We define *similarity* for input events as follows:

(SI-Net)
$$\frac{in\_erase_l(resp) = in\_erase_l(resp') \quad\quad * \in \{\mathsf{doc}, \mathsf{script}, \mathsf{xhr}\}}{\mathsf{net\_}*\mathsf{\_resp}(ncid, uwi, resp) \approx_l \mathsf{net\_}*\mathsf{\_resp}(ncid, uwi, resp')}$$

(SI-Refl)
$$i \approx_l i$$

Here, $in\_erase_l(resp)$ is obtained from $resp$ by erasing from $resp.set\_cookies$ the value of every cookie c such that $ck\_label(c) \not\sqsubseteq l$ with:

$$ck\_label(c) = \begin{cases} \mathsf{http}(d) & \text{if } c.domain = d \wedge c.httponly = \mathsf{true} \wedge c.secure = \mathsf{false} \\ \mathsf{https}(d) & \text{if } c.domain = d \wedge c.httponly = \mathsf{true} \wedge c.secure = \mathsf{true} \\ \bot & \text{otherwise} \end{cases}$$

Intuitively, $i \approx_l i'$ if and only if i and i' are syntactically equal, except for the cookies hidden to an attacker at level l (recall the previous informal overview).

We conclude by defining *visibility* for input events: that is an easy task, since we just stipulate that $visible_l(i)$ holds true for all security labels l and all input events i. In other words, this corresponds to stating that the occurrence of a given input event is never hidden to the attacker: indeed, the cookie flags described above are just intended to protect the *value* of the cookie.

Formal results. Let EFF denote the extended FF of Section 3.1: assuming the definitions of visibility and similarity for input and output events introduced above, we have our desired result.

Theorem 1 (Noninterference). *EFF is noninterferent.*

We refer the interested reader to Appendix A for an intuition about the coinductive technique adopted in the proof. The result is interesting and important in itself, however, as it provides a certified guarantee of the effectiveness of the Secure and $\mathsf{HttpOnly}$ flags as robust protection mechanisms for session cookies. Needless to say, the theorem does not say anything about the security of sessions in existing web applications, as that depends critically on the correct use of the cookie flags. In the next section, we analyze the actual deployment of such mechanisms in existing systems, and describe our approach to enforce their use at the client side to secure modern browsers.

4   Strengthening Session Security

4.1   Session Cookie Protection in Existing Systems

We start with an analysis of the actual adoption of the security flags in existing systems. To accomplish that, we conduct an analysis of the the top 1000 websites of Alexa: we first collect the cookies registered through the HTTP headers by these websites, and then apply a heuristic to isolate session cookies. The heuristic marks a cookie as a session cookie if it satisfies either of the following conditions:

1. the cookie name contains the strings 'sess' or 'sid';
2. the cookie value contains at least 10 characters and its index of coincidence[3] is below 0.04.

Our solution is consistent with previous proposals [17] and has been validated by a manual investigation on known websites. In particular, condition 1 is motivated by the observation that several web frameworks offer native support for cookie-based sessions and by default register session cookies with known names satisfying this condition. In addition, it appears that custom session identifiers tend to include the string 'sess' or 'sid' in their names as well. Condition 2, in turn, is dictated by the expected statistical properties of a robust session identifier, which is typically a long and random string. Clearly, there is no *a prori* guarantee of accuracy for our heuristic. As we will discuss, however, we have strong evidence that our survey is reliable enough (cf. Section 4.5).

4.2   The Need for a Client-Side Defense

Table 1 provides some statistics which highlight that the large majority of the session cookies we identified (71.35%) has no flag set: though this percentage may be partially biased by the adoption of a heuristic, it provides clear indications of a limited practical deployment of the available protection mechanisms. Further evidence will be provided by our field experiments.

Table 1. Statistics about cookie flags

HttpOnly	Secure	#cookies	percentage
yes	yes	32	2.81%
yes	no	284	24.96%
no	yes	10	0.88%
no	no	812	71.35%

Of the two flags, HttpOnly appears to be adopted much more widely than Secure. We conjecture two reasons for that: first, modern releases of major web frameworks (e.g., ASP) automatically set the HttpOnly flag (but not the Secure flag) for session cookies generated through the standard API; second, Secure cookies presuppose an HTTPS implementation, which is not available for all websites. We further investigate this point below.

---

[3] This is a statistical measure which can be effectively employed to understand how likely a given text was randomly generated [11].

Evaluating Client-Side Protection. Prior research has advocated the selective application of the HttpOnly flag to session cookies at the client side to reduce the attack surface against session hijacking [17,21]. We propose to push this idea further, by automatically flagging session cookies also as **Secure** and enforcing a redirection to HTTPS for supporting websites.

To get a better understanding about the practical implications of this approach, we conducted a simple experiment aimed at estimating the extent of the actual HTTPS deployment. We found that 192 out of the 443 websites registering at least one session cookie (43.34%) support HTTPS transparently, i.e., they can be successfully accessed simply by replacing http with https in their URL. (In this count we excluded a number of websites which automatically redirect HTTPS connections over HTTP.) We then observed that only 16 of these websites (8.33%) set the **Secure** flag for at least one session cookie. Remarkably, it turns out that 141 out of these 192 websites (73.44%) contain at least one HTTP link to the same domain hard-coded in their homepage, hence session cookies which are not marked **Secure** are at risk of being disclosed to a network attacker when navigating these websites.

4.3 Client-Side Protection with CookiExt

CookiExt is an extension for Google Chrome aimed at enforcing robust client-side protection for session cookies. We choose Chrome for our development because it provides a fairly powerful – yet simple to use – API for programming extensions: the same solution could be implemented in any other modern web browser.

Overview. At a high level, the behaviour of CookiExt can be summarized as follows: when the browser receives an HTTP(S) response, CookiExt inspects its headers, trying to identify the session cookies based on the heuristic discussed earlier. If a session cookie is found, CookiExt behaves as follows:

- if the response was sent over HTTPS, all the identified session cookies are marked **Secure** and HttpOnly;
- if the response was sent over HTTP, all the identified session cookies are erased from the HTTP headers.

In both cases, all subsequent requests to the website are automatically redirected over HTTPS. This simple picture, however, is significantly complicated by a number of issues which arise in practice and must be addressed to devise a usable implementation.

Supporting "mixed" Websites. Mixed websites are websites which support HTTPS but make some of their contents available only on HTTP. This website structure is often adopted by e-commerce sites, which offer access to their private areas over HTTPS, but then make their catalogs available only on HTTP. These cases are problematic, as enforcing a redirection over HTTPS for the HTTP portion of the website would make the latter unavailable. Similarly, assuming to be able to detect the absence of HTTPS support for some links, even the adoption of a fallback to HTTP would eventually break the user's session: in fact, since session cookies are by default marked Secure by our extension, they will not be sent to the HTTP portion of the website.

We therefore adopt the following, more elaborate solution. When CookiExt forces a redirection over HTTPS, we implement a check to detect possible failures (see below). If the redirection cannot be performed, we enforce a fallback to HTTP, distinguishing two cases: if the failure arises from the request of a page, we extend the set of the cookies attached to the request with all the cookies which have been made Secure by CookiExt, but were not originally marked Secure by the server; if the browser instead is trying to retrieve a sub-resource, like an image or a script, we leave the set of cookies attached to the request unchanged. This way we confine any deviation from the browser behaviour expected by the server only to sub-resources, which typically do not require authenticated access. Clearly, transmitting in clear the session cookies identified by the heuristic exposes the client to a risk. However, this approach offers an interesting compromise between usability and security: in fact, the user's navigation will not break the session, since page requests will always include session cookies, but at the same time the attack surface for network attackers will be significantly reduced, since retrieving an image over HTTP inside an HTTPS page will not leak any session cookie.

Checking HTTPS Support. As we said, CookiExt could try to enforce a redirection from HTTP to HTTPS also for websites supporting only HTTP access. The Chrome API already allows one to detect a number of network connection errors which may arise when HTTPS is not supported; however, some of these alerts are only triggered after a significant delay, which may negatively affect usability.

Our choice is to set a relatively small timeout every time CookiExt forces an HTTPS redirection: if no response is received before the timeout expires, the extension fallbacks to HTTP. To prevent a network attacker from tapping with outgoing HTTPS connections and disabling our client-side defense, CookiExt keeps track of all the pages for which a successful redirection from HTTP to HTTPS has been performed in the past, and notifies the user in case of an unexpected lack of HTTPS support possibly due to malicious network activities.

4.4   Noninterference in Theory and in Practice

Careful readers will argue that our non-interference result (Theorem 1) predicates on (a Coq model of) a standard web browser rather than on a web browser extended with CookiExt, and consequently provides no information about the soundness of CookiExt. The gap is only apparent, however, as CookiExt does not really alter the browser behaviour, but rather *activates* existing protection mechanisms available in standard web browsers. Indeed, one may view CookiExt just as a filter that applies the desired flagging to all inputs, *de facto* enforcing the similarity condition on the input streams that constitutes the hypothesis of the non-interference definition. The only discrepancy determined by CookiExt arises from the fallback mechanism, which may end up sending over HTTP cookies that the extension promoted to Secure. While this effect has no counterpart in standard browsers, the gap is again harmless, as all such cookies can be assimilated to HttpOnly cookies, for which confidentiality is guaranteed against web attackers.

4.5   Experiments

The effectiveness of CookiExt critically depends on the accuracy of the heuristic for session cookie detection. On the one hand, false negatives lead to failures at protecting the session cookies of vulnerable websites. On the other hand, false positives may hinder the usability of the browser (though they do not cause any security flaw). We now report on our experiments to evaluate both these aspects.

Security Evaluation. To understand the practical impact of false negatives, we analyze again our survey of websites and isolate the cookies flagged Secure or HttpOnly which are not identified as session cookies by the heuristic: the intuition here is that cookies which are explicitly protected by web developers are likely to contain session information and we deem them as potential false negatives. As it turns out, only 37 of the 1153 cookies ignored by our heuristic (3.21%) have at least one security flag set: in addition, 8 of these 37 cookies are already flagged Secure and HttpOnly, hence missing them is completely harmless. We then carried out a manual review of the remaining potential false negatives, which showed that none of the 29 cookies left is a real session identifier. We performed this check by authenticating onto the private areas of the websites registering one of these cookies, and then erasing all the cookies identified by our heuristic: a logout from the website implies that all the real session cookies have been successfully recognized.

Usability Tests. The only way to understand the practical impact of the false positives is by testing and hands-on experience with the extension. We performed an empirical evaluation by having a small set of users install CookiExt on their browsers and navigate the Web, trying to find out usability issues and general limitations while performing standard operations on websites where they own a personal account. The feedback by the users was very important to refine the original implementation and make it work in practice: with our latest prototype, no major complaint was reported at the time of writing, even though some users have been noticing a slight performance degradation when activating CookiExt.

Manual Investigation. We carried out a manual investigation on the top 20 websites from the Alexa ranking where we own a personal account. For all the websites we performed three different experiments:

1. Detecting session cookies. We authenticate to the private area of the website and we delete from the browser all the cookies which have been marked as session cookies by our extension: a logout implies that all the real session cookies have been identified by our heuristic. In all cases, our heuristic over-approximated correctly the real set of session cookies;

2. Preserving usability. We navigate the website as deep as possible, trying to identify visible usability issues. Our most serious concern was about the web session being broken by the security policy applied by CookiExt, but this never happened in practice. From our experience, usability crucially hinges on our choice of discriminating the behaviour of CookiExt based on the request type. Occasionally, we noticed that some images are not loaded when our extension is activated: it seems this typically happens with third-party advertisement, since the choice of the contents to deliver to the browser likely depend on some tracking cookies which are stripped off by CookiExt;

3. Evaluating protection. We navigate the website and we log any redirection attempt from HTTP to HTTPS when navigating to an internal web page: we identified 36 requests overall, among which 23 were successfully redirected to HTTPS; we manually verified that 11 of the 13 remaining links do not support HTTPS.

We remark that, though promising, our extension is still a prototype under active development: we are currently performing a larger scale evaluation and implementing a number of practical improvements.

5   Related Work

Browser-Side Protection Mechanisms. The idea of enforcing security browser-side is certainly not new. Below, we focus on a detailed comparison with the works which share direct similarities with our present proposal. Other approaches exist as well [15,14,18,19,20], but the relationships with ours are loose.

SessionShield [17] is a lightweight protection mechanism against session hijacking. SessionShield acts as a proxy between the browser and the network: incoming session cookies are stripped out from HTTP headers and stored in an external database; on later HTTP requests, the database is queried using the domain of the request as the key, and all the retrieved session cookies are attached to the outgoing request. We find the design of SessionShield very competent and we borrowed the idea of relying on a heuristic to identify session cookies in our implementation. On the other hand, SessionShield does not enforce any protection against network attacks and does not support HTTPS, since it is deployed as a stand-alone personal proxy external to the browser.

The idea of identifying session cookies through a heuristic and selectively applying the HttpOnly flag to them has also been advocated in Zan [21], a browser-based solution aimed at protecting legacy web applications against different attacks. Similarly to SessionShield, Zan does not implement any protection mechanism against network attackers.

Another particularly relevant client-side defense is HTTPS Everywhere [22]. This is a browser extension which enforces communication with many major websites to happen over HTTPS. The tool also offers support for setting the Secure flag of known session cookies at the client side. Unfortunately, HTTPS Everywhere does not enforce any protection against XSS attacks, hence it does not implement complete safeguards for session cookies. Moreover, the tool relies on a white-list of known websites both for redirecting network traffic over HTTPS and to identify session cookies to be set as Secure, an approach which does not scale in practice and fails at protecting websites not included in the white-list. Similar design choices and limitations apply to ForceHTTPS [13], a proposal aimed at protecting high-security websites from network attacks.

Formal Methods for Web Security. The importance of applying formal techniques to web security has been first recognised in a seminal paper by Akhawe *et al.* [2]. The work proposes a rigorous formalization of standard web concepts (browsers, servers, network messages...), a clear threat model and a precise specification of its security goals. The model is implemented in Alloy and applied to several case studies, exposing concrete attacks on web security mechanisms.

A more recent research paper by Bansal *et al.* [3] introduces WebSpi, a ProVerif library which provides an applied pi-calculus encoding of a number of web features, including browsers, servers and a configurable threat model. The authors rely on the WebSpi library to perform an unbounded verification of several configurations of the OAuth authorization protocol through ProVerif, identifying some previously unknown attacks on authentication.

Reactive Noninterference. The theory of reactive noninterference has been first developed by Bohannon *et al.* [7]. Aaron Bohannon's doctoral dissertation [5] provides a mechanized proof of noninterference for a Coq implementation of the original Featherweight Firefox model [6] extended with a number of dynamic checks aimed at preventing information leakage. In the present work we leverage the existing proof architecture to carry out our formal development, with the notable differences and extensions discussed in Section 3 and Appendix A.

Independently from Bohannon's work, Bielova *et al.* [4] proposed an extension of the Featherweight Firefox model to enforce reactive noninterference through a dynamic technique known as secure multi-execution. In later work, De Groef *et al.* [12] built on this approach to develop FlowFox, a full-fledged web browser implementing fine-grained information flow control.

6   Conclusion

We have provided a formal view of web session security in terms of reactive noninterference and we showed that the protection mechanisms available in modern web browsers are effective at enforcing this notion. On the other hand, our practical experience highlighted that many web developers still fail at adequately protecting session cookies, hence we proposed CookiExt, a client-side solution aimed at taming existing security flaws. We find preliminary experiences with our tool to be fairly satisfactory.

We imagine different directions for future work. First, we would like to further refine our formal model, to include additional concrete details which were initially left out from our study for the sake of simplicity. Moreover, we plan to combine the current heuristic for session cookie detection with a learning algorithm, to improve its accuracy by analysing the navigation behaviour.

Finally, we remark that both our theory and implementation just focus on the *confidentiality* of session cookies, which is a necessary precondition for thwarting the risk of session hijacking. However, several serious security threats against web sessions do not follow by confidentiality violations: for instance, classic CSRF vulnerabilities should rather be interpreted in terms of attacks on integrity. In a recently submitted paper we consider a much stronger definition of web session security and we discuss its browser-side enforcement [8].

References

1. The Coq proof assistant, http://coq.inria.fr/
2. Akhawe, D., Barth, A., Lam, P.E., Mitchell, J.C., Song, D.: Towards a formal foundation of web security. In: IEEE Computer Security Foundations Symposium (CSF), pp. 290–304 (2010)
3. Bansal, C., Bhargavan, K., Maffeis, S.: Discovering concrete attacks on website authorization by formal analysis. In: IEEE Computer Security Foundations Symposium (CSF), pp. 247–262 (2012)
4. Bielova, N., Devriese, D., Massacci, F., Piessens, F.: Reactive non-interference for a browser model. In: IEEE International Conference on Network and System Security (NSS), pp. 97–104 (2011)
5. Bohannon, A.: Foundations of webscript security. PhD thesis, University of Pennsylvania (2012)
6. Bohannon, A., Pierce, B.C.: Featherweight Firefox: formalizing the core of a web browser. In: USENIX Conference on Web Application Development (WebApps), Berkeley, CA, USA, pp. 1–12. USENIX Association (2010)
7. Bohannon, A., Pierce, B.C., Sjöberg, V., Weirich, S., Zdancewic, S.: Reactive non-interference. In: ACM Conference on Computer and Communications Security (CCS), pp. 79–90 (2009)
8. Bugliesi, M., Calzavara, S., Focardi, R., Tempesta, M., Khan, W.: Formalizing and enforcing web session integrity (submitted)
9. Dacosta, I., Chakradeo, S., Ahamad, M., Traynor, P.: One-time cookies: Preventing session hijacking attacks with stateless authentication tokens. ACM Transactions on Internet Technology 12(1), 1 (2012)
10. Fogie, S., Grossman, J., Hansen, R., Rager, A., Petkov, P.D.: XSS Attacks: Cross Site Scripting Exploits and Defense. Syngress Publishing (2007)
11. Friedman, W.F.: The index of coincidence and its applications to cryptanalysis. Cryptographic Series (1922)
12. Groef, W.D., Devriese, D., Nikiforakis, N., Piessens, F.: FlowFox: a web browser with flexible and precise information flow control. In: ACM Conference on Computer and Communications Security (CCS), pp. 748–759 (2012)

13. Jackson, C., Barth, A.: Forcehttps: protecting high-security web sites from network attacks. In: International Conference on World Wide Web (WWW), pp. 525–534 (2008)
14. Johns, M., Winter, J.: RequestRodeo: client side protection against session riding. In: Proceedings of the OWASP Europe Conference, pp. 5–17 (2006)
15. Kirda, E., Krügel, C., Vigna, G., Jovanovic, N.: Noxes: a client-side solution for mitigating cross-site scripting attacks. In: ACM Symposium on Applied Computing (SAC), pp. 330–337 (2006)
16. Liu, A.X., Kovacs, J.M., Gouda, M.G.: A secure cookie scheme. Computer Networks 56(6), 1723–1730 (2012)
17. Nikiforakis, N., Meert, W., Younan, Y., Johns, M., Joosen, W.: SessionShield: Lightweight protection against session hijacking. In: Erlingsson, Ú., Wieringa, R., Zannone, N. (eds.) ESSoS 2011. LNCS, vol. 6542, pp. 87–100. Springer, Heidelberg (2011)
18. Nikiforakis, N., Younan, Y., Joosen, W.: HProxy: Client-side detection of SSL stripping attacks. In: Kreibich, C., Jahnke, M. (eds.) DIMVA 2010. LNCS, vol. 6201, pp. 200–218. Springer, Heidelberg (2010)
19. De Ryck, P., Desmet, L., Joosen, W., Piessens, F.: Automatic and precise client-side protection against CSRF attacks. In: Atluri, V., Diaz, C. (eds.) ESORICS 2011. LNCS, vol. 6879, pp. 100–116. Springer, Heidelberg (2011)
20. De Ryck, P., Nikiforakis, N., Desmet, L., Piessens, F., Joosen, W.: Serene: Self-reliant client-side protection against session fixation. In: Göschka, K.M., Haridi, S. (eds.) DAIS 2012. LNCS, vol. 7272, pp. 59–72. Springer, Heidelberg (2012)
21. Tang, S., Dautenhahn, N., King, S.T.: Fortifying web-based applications automatically. In: ACM Conference on Computer and Communications Security (CCS), pp. 615–626 (2011)
22. Tor Project and the Electronic Frontier Foundation. HTTPS Everywhere. Available for download at, https://www.eff.org/https-everywhere

A   Noninterference Proof

Following previous work [7,4], we prove our main result through an unwinding lemma, which provides a coinductive proof technique for reactive noninterference. However, we depart from previous proposals by developing a variant of the existing unwinding lemma based on a *lockstep* unwinding relation.

Definition 6 (Lockstep Unwinding Relation). *We define a lockstep unwinding relation on a reactive system as a label-indexed family of binary relations on states (written $\simeq_l$) with the following properties:*

1. *if $Q \simeq_l Q'$, then $Q' \simeq_l Q$;*
2. *if $C \simeq_l C'$ and $C \xrightarrow{i} P$ and $C' \xrightarrow{i'} P'$ and $i \approx_l i'$ and $visible_l(i)$ and $visible_l(i')$, then $P \simeq_l P'$;*
3. *if $C \simeq_l C'$ and $\neg visible_l(i)$ and $C \xrightarrow{i} P$, then $P \simeq_l C'$;*
4. *if $P \simeq_l C$ and $P \xrightarrow{o} Q$, then $\neg visible_l(o)$ and $Q \simeq_l C$;*
5. *if $P \simeq_l P'$, then for any o, o', Q, Q' such that $P \xrightarrow{o} Q$ and $P' \xrightarrow{o'} Q'$ we have $Q \simeq_l Q'$, provided that either (i) $o \approx_l o'$; or (ii) $\neg visible_l(o)$ and $\neg visible_l(o')$.*

With respect to previous proposals, the main difference is in clause 5, where we require two related producer states to proceed in a lockstep fashion, even when they emit invisible output events. We can show that exhibiting a lockstep unwinding relation on the initial state of a reactive system is enough to prove noninterference.

Lemma 1 (Unwinding). *If $Q \simeq_l Q$ for all l, then Q is noninterferent.*

Proof. We show by coinduction that $Q \simeq_l Q'$ implies $Q \sim_l Q'$ for all l, where $\sim_l$ is an unwinding relation according to the definition in [5]. Then, the result follows by the main theorem therein, showing that, if $Q \sim_l Q$ for all l, then Q is noninterferent.

By relying on a lockstep unwinding relation rather than on a standard unwinding relation, we can dramatically simplify the definition of the witness required by our proof technique and the proof itself, as we discuss below.

We can finally give a solid intuition about our main result. The browser state b in Featherweight Firefox is represented by a tuple, which contains several data structures representing open windows, loaded pages, cookies, open network connections and a bunch of additional information needed for the browser to operate. We identify the set of *consumer states* with the space state generated by instantiating the set of these data structures in all possible ways. We then define *producer states* by pairing a consumer state b with a task list t: this list keeps track of the script expressions that the browser must evaluate before it can accept another input. State transitions are defined by the FF implementation: intuitively, the browser starts its execution in a consumer state and each kind of input fed to it will initialize the task list in a different way. Processing the task list moves the browser across producer states (possibly adding new tasks): when the task list is empty, the browser moves back to a consumer state.

To prove noninterference, we define our candidate lockstep unwinding relation $\simeq_l^B$ as follows:

$$
\begin{array}{cc}
\text{(B-Cons)} & \text{(B-Prod)} \\[4pt]
\dfrac{erase_l(b) = erase_l(b')}{b \simeq_l^B b'} & \dfrac{b \simeq_l^B b'}{(b,t) \simeq_l^B (b',t)}
\end{array}
$$

where $erase_l(b)$ is obtained from b by erasing from its cookie store the value of every cookie c with $ck\_label(c) \not\sqsubseteq l$. We then show that $\simeq_l^B$ is indeed a lockstep unwinding relation and that $b_{init} \simeq_l^B b_{init}$ for all l, where b_{init} is the initial state of the Featherweight Firefox model. By Lemma 1, this implies that the browser model is noninterferent. As a technical note, we point out that $\simeq_l^B$ is not itself an unwinding relation according to the definition in [5]. On the other hand, it is a lockstep unwinding relation, which is enough for our present needs.

Security Testing of GSM Implementations

Fabian van den Broek[1], Brinio Hond[2], and Arturo Cedillo Torres[2]

[1] Institute for Computing and Information Sciences,
Radboud University Nijmegen, The Netherlands
f.vandenbroek@cs.ru.nl
[2] KPMG
{hond.brinio,cedillotorres.arturo}@kpmg.nl

Abstract. Right after its introduction, GSM security was reviewed in a mostly theoretical way, uncovering some major security issues. However, the costs and complexity of the required hardware prohibited most people from exploiting these weaknesses in practice and GSM became one of the most successful technologies ever introduced. Now there is an enormous amount of mobile enabled equipment out there in the wild, which not only have exploitable weaknesses following from the GSM specifications, but also run implementations which were never security tested. Due to the introduction of cheap hardware and available open-source software, GSM found itself under renewed scrutiny in recent years. Practical security research such as fuzzing is now a possibility.

This paper gives an overview on the current state of fuzzing research and discusses our efforts and results in fuzzing parts of the extensive GSM protocol. The protocol is described in hundreds of large PDF documents and contains many layers and many, often archaic, options. It is, in short, a prime target for fuzzing. We focus on two parts of GSM: SMS messages and CBS broadcast messages.

1 Introduction

GSM saw its first deployment in 1991 in Finland and from there grew out to become one of the dominant technologies. GSM can be considered old technology, since there are numerous newer technologies in the GSM family, such as UMTS and LTE, which provide better bandwidth and possibilities for data transfer. However, that does not mean GSM is no longer a critical infrastructure. As of 2013, approximately 7 billion SIM cards are active worldwide, offering subscription services to the GSM family of networks for around 3.4 billion unique subscribers [1]. Even though these subscriptions are partly for other networks, GSM is nearly always the base subscription and almost all equipment supports GSM (or GPRS for mobile Internet equipment). Furthermore, GSM/GPRS coverage is far more extensive than the coverage of the newer protocols and GSM uses less power and is more efficient for voice calls. Also, a lot of machine-to-machine communication relies on GSM/GPRS, such as certain smart meters and traffic lights in South Africa [2,3] as well as railway systems in the European Union [4]. All this has prompted providers to speculate that newer protocols such as

J. Jürjens, F. Piessens, and N. Bielova (Eds.): ESSoS 2014, LNCS 8364, pp. 179–195, 2014.
© Springer International Publishing Switzerland 2014

LTE will replace their direct predecessor (UMTS), but will still run alongside an active GSM/GPRS network [5]. So, for the foreseeable future, GSM is here to stay. When GSM was first deployed there was some security research, which mostly focused on the specifications and the reverse engineering of the secret and proprietary encryption algorithm [6]. Several weaknesses in GSM where quickly identified, though practical exploits of these weaknesses proved complicated because of all the signal processing involved. This changed around 2010 with the arrival of cheap hardware [7] and open-source software [8] which provided easy access to the GSM spectrum. This immediately led to some high profile attacks, such as the release of the Time-Memory Trade-Off tables for breaking GSM's standard encryption [9,10].

With this new hardware and software it is possible to run your own GSM cell tower to which real phones will connect, since in GSM the network does not authenticate itself to the phones. This opens up the possibility to verify the implementations of the GSM stack of phones by the technique known as fuzzing. Fuzzing has been used a lot to find security holes on Internet equipment. Thanks to low level access offered by Ethernet cards it was easy to simply try out all kinds of possible messages, mostly those just outside of the specifications, and see what happens when these are received by network equipment. Fuzzing mobile phones has mostly happened in the hackers scene of security research, with few academic publications.

Naturally, there are many interfaces in mobile phones which can be fuzzed. Just think of every type of input that a phone can receive, such as WiFi, Bluetooth, NFC, installed apps or the SIM interface. All of these inputs can be interesting input vectors for fuzz testing. We focused on fuzzing the GSM baseband stack. This is the part of the phone which handles all the GSM traffic. It is available in every phone, implements a hugely complicated standard and is remotely accessible over the air, which could easily lead to dangerous attacks.

The GSM system comprises many entities, such as the mobile phones and cell towers, but also many more back-end components. Our fuzzing research only focuses on mobile phones. Naturally, fuzzing the network components of a GSM network can have a much larger impact. However, availability of commercially used network components that are not currently running inside an operational GSM network is very limited. Thus we limited ourselves to the readily available mobile phones. In this paper we discuss our efforts and results in fuzzing two specific parts of the GSM specification: SMS messages and CBS messages.

The well-known Short Message Service (SMS) was added shortly after the initial release of GSM and the first SMS message was sent in 1992 [1]. The first version of SMS allowed the exchange of short text messages between GSM users, but SMS has gone a long way since then. Not only can SMS be used to exchange text messages, but nowadays also pictures, sounds and many other types of data can be sent over the SMS. The current SMS standards also allow segmentation of messages that are too long to fit into a single message, enabling users to transmit much longer messages. The current SMS specification is found in [11,12].

The lesser-known Public Warning System (PWS) actually started out as the Cell Broadcast Service (CBS), which was developed in parallel to the SMS service as a response of mobile developers to the competing paging services being offered in 1990. It allows providers to broadcast messages to all phones currently connected to a certain cell, i.e. all phones connected to a single transceiver on a cell tower. The original business case was to provide news, weather and traffic information to mobile users, though this never found any wide spread popularity. This lead to both mobile network operators and mobile developers neglecting the implementation of the service in their equipment. However, this service has been gaining importance in the last years, because it can be an ideal method for governments to broadcast information in the event of an emergency to all phones in the vicinity. Several countries define and implement their own warning system that rely on the CBS to deliver emergency information. Due to the diversity of technical specifications of each warning system, ETSI with the aid of the 3GPP consortium developed a standardized system known as the Public Warning System (PWS). The initial goal of the PWS was to introduce a standard emergency and warning communication infrastructure, as well as specific technical requirements for mobile phones within the European Union to receive these emergency messages. Due to its standardized nature this system and its accompanying protocols can now be implemented worldwide. This allows roaming users to receive broadcast messages no matter what their location is, as long as they are in a GSM coverage area.

Structure of this Paper. In Section 2 we discuss the basics of the GSM air interface and provide an introduction into the SMS and CBS protocols. We then discuss fuzzing in general and the specifics of fuzzing mobile phones in Section 3. Section 4 describes our own fuzzing research, together with the practical details and results. It is here that we also discuss the related work, for comparison and to attempt to provide an overview of the fuzzing research into GSM up to this point. Finally, Section 5 presents the conclusions and ideas for future work.

2 GSM

The GSM baseband stack is usually described in three layers, where the third layer is again subdivided, as is shown in Figure 1. The bottom two layers of the GSM stack show similarities with the OSI model. The first layer, the physical layer, creates a set of logical channels through time division on already divided frequencies. These channels can be used by higher layer functions for many different tasks, as uplink (mobile phone to cell tower), downlink (cell tower to mobile phone) or broadcast (cell tower to all connected phones) communication. These channels can either be a traffic channel, or one of a multitude of control channels. Most control data is transmitted in 184 bit frames which are split up into 4 bursts. These bursts are modulated and transmitted by radio waves.

The signaling protocol used on the second layer, the data link layer, is called LAPDm. The data link layer (and higher layers) is only defined for the signaling channels, not for the speech channels. This is because speech bursts contain no further headers or other meta information, only speech data; during a phone conversation, the traffic on the dedicated speech channels needs no meta information in order to be reconstructed correctly at the receiving end. The LAPDm protocol can provide positive acknowledgement, error protection through retransmission, and flow control.

The third layer is where the match with the OSI model stops. The third layer is subdivided into three layers, of which the last (highest) one is again subdivided into several protocols:

1. Radio Resource management (RR); this concerns the configuration of the logical and physical channels on the air-interface;
2. Mobility Management (MM); for subscriber authentication and maintaining the geographical location of subscribers;
3. Connection management (CM); consists of several sublayers itself, such as:
   (a) Supplementary Services (SS); managing all kinds of extra services that are not connected to the core functionality of GSM;
   (b) Short Message Service (SMS); the handling of the SMS messages;
   (c) Call Control (CC); creating and ending telephone calls;
   (d) Locations Services (LCS); location based services for both the user and the provider;

Layer 3 frames consist of a 2 byte header followed by 0 or more Information Elements (IEs). These IEs can be of several different types: T, V, TV, LV and TLV, where the letters T, L and V denote the presence of a Type, Length and Value field respectively. The type field is always present in non-mandatory IEs. Interesting from a fuzzing perspective are those IEs that contain a length field, LV and TLV, even though they are specified as having a standard length, because these are typical places where a programmer might make a mistake in handling data of non-standard length.

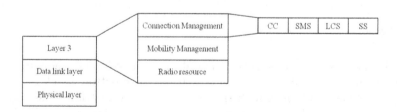

Fig. 1. The layers of GSM

We only fuzz on the third layer of the protocol stack, since this is more likely to trigger observable bugs than fuzzing on the first two layers. That is not to say that the lower layers of the protocol will likely contain less, or less nasty bugs, they are simply harder to observe.

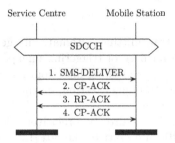

Fig. 2. Message sequence chart of delivering an SMS to a mobile phone

SMS. Before messages can be sent on the SMS sublayer the cell tower needs to notify the mobile phone of an incoming message and set up the channel (Standalone Dedicated Control Channel or SDCCH) with the mobile phone. The delivery of an SMS message then requires four messages exchanged on the SMS sublayer using the SDCCH, as shown in Figure 2. The first message is the SMS-DELIVER message sent from the network to the phone. This message contains the actual content (user data) with an optional User Data Header (UDH) and mandatory Transfer Protocol (TP), Relay Protocol (RP) and Connection Protocol (CP) headers. The phone first parses the CP header and verifies it. If it is valid the MS returns a CP-ACK message, otherwise it returns a CP-ERROR message and releases the connection. If the CP header was correct the MS continues by verifying the RP header and checking if the phone has enough memory to store the message. If either of those checks fails it returns an RP-ERROR and releases the connection. If both checks succeed the MS returns an RP-ACK with a CP header. The final message is sent by the network when the RP-ACK passes the checks for the CP header.

A schematic overview of a correct SMS-DELIVER message can be found in Figure 3, where Figure 3(b) shows the fields we fuzzed of the SMS-DELIVER message. Of the RP-ACK message we fuzzed practically all header fields.

Public Warning System. For CBS messages no specific traffic channels need to be set up for mobile phones to receive the transmission. The messages are transmitted on the broadcast channel, a specific channel to which all phones always listen to see if they are still in the same cell. So even if a cell is overloaded with regular voice or data traffic, broadcast messages can still be sent to mobile phones. This very feature is what makes them interesting for emergency messages in the first place. Japan's tsunami warning system and the European emergency broadcast (EU-Alert) are examples of implementations of the PWS.

A CBS message is first announced on a broadcast channel and then transmitted in four frames. A schematic representation of a CBS message is shown in Figure 4 where all the fields we fuzzed are shown in grey.

3   Fuzzing

Fuzzing is the process of transmitting automatically generated, uncommon inputs to a target with the purpose of triggering unexpected behavior. This unexpected behavior is typically something like program crashes or failing built-in code assertions. In contrast to human testing, fuzzing can be largely automated and as a result can find (security) errors that likely won't be triggered during normal use or testing. Fuzzing is already a relatively old testing technique, dating from the end of the 70s and start of the 80s [13]. Fuzzing has evolved over the years into several variants:

1. Plain fuzzing,
2. Protocol fuzzing, and
3. State-based fuzzing.

Plain fuzzing is the original idea behind fuzzing: simply generating lots of test cases, often with random data, and feed this to the program you are testing. These test cases are usually made by mutating correct inputs and is used to test the error-handling routines. It is a highly portable way of fuzzing, but also provides very little assurance on code coverage.

In protocol fuzzing the test cases are generated based on the specifications, especially on specifications of packet formats. Here the fuzzer will try to choose specific test cases which are likely to provide the largest code coverage based on its knowledge of the specifications. Typically, fuzzers will look at each field which can contain more values than are allowed by the specifications and generate test cases with values on the corner cases, values just over the corner cases and some values way out of the range of what is allowed. This is also called "partition fuzz testing", since the possible inputs for a field are partitioned (e.g. a half byte which represents an ID and allows for the values 1-12, would give three partitions: 0, 1 to 12 and 13 to 15). These are often not partitions in the mathematical sense, as they need not be disjoint, but the union of all partitions usually do span the entire input space. Note that although these fuzzers are named "protocol fuzzers" and are in fact mostly used to test protocols, they can also be used to test non-protocol implementations, as long as the input has a format to which it should conform.

The first two fuzz variants discussed here try to find errors by changing the content of individual messages. But there is another part that can often be fuzzed: the state machine. Most protocols have some sort of set sequence in which messages are exchanged and which messages are expected at any one time is tracked in a state machine. When the wrong message is sent at some point in time and still accepted by the implementation it shows a problem with its state machine. The impact of this is hard to estimate, because it depends on what states can be skipped, but for some protocols it might allow one to bypass authentication steps, posing a serious security risk. State-based fuzzers not only change the content, but also the sequence of messages.

Whichever fuzzing approach is used, fuzzing will usually follow three distinct phases:

1. Generating the test cases,
2. Transmitting the test cases, and
3. Observing the behavior.

The fuzzing approaches discussed above concern the first phase. On traditional computer networks the second phase is trivial. Also, observing the effects after transmitting the fuzz tests is usually easier on traditional computers than on GSM phones, because there is often the possibility of running a debugger, or simply looking for familiar error messages. Fuzzing GSM implementations on baseband chips has many of the same problems as the fuzzing of embedded systems; one cannot easily observe the effect of fuzzed inputs [14].

3.1 Fuzzing GSM Phones

There are many different GSM-enabled mobile phones. Mobile phones started out with just the ability to make and receive voice calls, but nowadays phones are available that have a wide range of features and possible connections. It is important to realize that the current market contains a wide variety of GSM-enabled devices, not only of different make and model, but also internally: GSM phones can consist of a single processor, which runs the GSM protocol stack and a very limited OS for the user interface, these are typically cheaper or older GSM phone models. The chip running the GSM protocol stack is referred to as the *baseband chip* [15]. More complicated phones run their OS on a separate general purpose processor, named the *application processor*. Both processors can communicate through a variety of protocols, where the application processor uses the baseband processor like a modem. Most modern phones combine the application and baseband processor in a single SoC (System on a Chip).

Fuzzing mobile phones is challenging compared to e.g. fuzzing network cards, mainly because it is hard to observe undefined behavior. Most phones are closed devices without any debugging tools, so it is impossible to, for instance, look at the memory during operation. Also most phones run closed source software. This makes it harder to predict where errors will occur. Even Android phones use closed software libraries for low level communication with the baseband chip and if the baseband chip has its own memory a debugger on a rooted Android phone will provide little extra help. Phones usually have limited interaction possibilities, which makes observation a time consuming effort. Generally there are few alternatives to simply using the phone after a fuzz message and observing whether it shows any undefined behavior, which can lead to false positives. This means internal errors that do not directly lead to observable undefined behavior, may go unnoticed.

The fuzzed messages need to be introduced to the target phones. This can either be done by transmitting them as actual GSM messages to the phones, or by directly inserting them in the phones, for instance by inserting them on the wire between the baseband and application chip. The latter option is cumbersome to use on many different phones and much harder on modern phones with a single SoC, but this was the only available option when open source GSM networks were not available [16].

For transmitting the messages over actual networks there are several options:

1. You could use a running GSM network, either because you happen to have access to commercial GSM network equipment, or by transmitting fuzzed messages from a modified phone to a target phone over the normal network.
2. A more feasible solution is to change existing GSM equipment, such as a femtocell [17] for transmitting fuzz messages, though the success of this method will depend on the success in breaking the femtocell security.
3. Finally there are several open-source projects that allow you to set-up your own GSM network.

Using the existing network (option 1) severely limits the fields you can fuzz and the network operator could change or filter our messages. Adapting a femtocell to do the fuzzing (option 2) could prove unsuccessful, so we chose the third option. The most important of the open-source projects are OpenBTS [8] and OpenBSC [18]. Both systems run on most ordinary PCs and require extra hardware for transceiving GSM signals.

OpenBTS is based on the GNU Radio project [19] and is designed to work with the USRP (Universal Software Radio Peripheral), a generic and programmable hardware radio component. The USRP can be modified through the use of daughterboards for specific applications and frequencies. Several versions of the USRP are currently available and a typical setup for a local GSM network costs around $1500,- [7].

OpenBSC runs a basestation controller and therefore interfaces with an actual basestation in order to work. OpenBSC started out as a controller for the Siemens BS11, an actual commercial cell tower of which a small batch became available on eBay, and the nanoBTS from ip.access, a corporate solution miniature cell tower. Both cell towers are hard to obtain, so OpenBSC will now also work with a new project, OsmoBTS [20], which in turn implements a cell tower on several devices such as the custom made sysmoBTS and even, experimentally, two modified mobile phones.

The OsmoBTS project was not yet available when we started our research, which led us to choose the OpenBTS option, for full control of the GSM air link.

4   Our Fuzzing

For GSM all the specifications are openly available, but implementations of the baseband stacks are not. Examining the specifications led to the conclusion that although GSM is a very complicated protocol, there are actually very few state changes in the baseband stacks. This is why we mostly resorted to protocol fuzzing, as will be described in Section 4.1 and 4.2. We attempted some state-based fuzzing on the SMS sublayer by both sending a correct message when it was not expected by the phone and sending a correct message when a different message was expected by the phone. This only showed unexpected results for one phone, the Sony-Ericsson T630, which accepted confirmations of unsent SMS messages, but which did not lead to any exploitable results.

There are several open-source protocol fuzzing frameworks available [21]. However, these frameworks are not made to be used with cell phones. Especially the target monitoring aspects generally work on network interfaces and virtual machines, while we have separate devices connected over a (custom) radio interface. This makes automatic target monitoring with one of these tools impossible. We did end up using one fuzzer, Sulley [22], as the basis for our fuzzer.

4.1   How Do We Fuzz?

For this research we made our own fuzzer GSMFuzz, for the generation of the fuzz messages. It is a fuzzer, with features designed specifically for GSM, but which nonetheless can be used for other protocols as well. The fuzzer is written in Python (version 2.6) and loosely based on Sulley[22], an open-source fuzzing framework. It has the following features added:

- Fuzzing of bit positions within a byte;
- Partition fuzz testing of special fields (type, length), resulting in few cases with maximum impact;
- Innate support for the eight different GSM Layer 3 IEs;
- Length fields can count octets, septets or half-octets (often used in GSM);
- Hexadecimal output of fuzz cases to a file, which can be used directly in our extended version of OpenBTS.

GSMfuzz itself is just over 900 lines of code (excluding white space). Besides the source code of the program itself we created 34 files with input to mutate valid messages. The input files are 3601 lines in total (excluding white space and comments).

Figure 3(b) shows the fields we fuzzed in the SMS-DELIVER message and Figure 4 shows the fields we fuzzed in the CBS message.

For the transmission of the fuzzed messages we used the open-source OpenBTS software together with a USRP-1 (where the internal clock was replaced with the more precise Fairwaves ClockTamer-1.2) and a collection of (a WBX and two RFX1800) daughterboards. Combining this with two Ettus LP0926 900 MHz to 2.6 GHz antennas yielded a setup of around 1500 €.

We tuned the software to only allow a specific set of SIM cards to connect, but this did not prevent several phones in the surroundings to still connect to our cell. This already shows errors in how phones handle connecting to cell towers. Since we did not want to unintentionally harm the phones of our colleagues, we made a Faraday cage around the whole setup, using chicken wire. With a maze size of 12.5mm, which is smaller than ten times the wavelength of GSM signals on 1800MHz, we managed to keep our GSM broadcasts contained.[1]

---

[1] There was some leakage through the power cord, but not enough to get phones outside of the cage to connect.

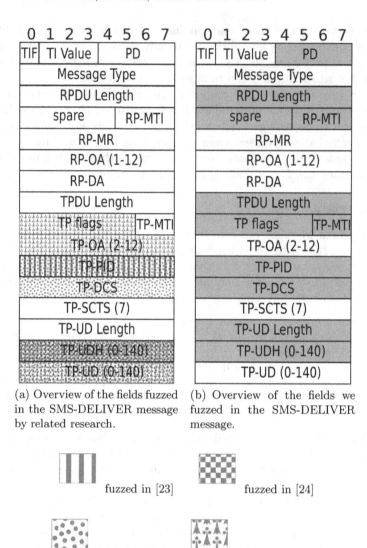

(a) Overview of the fields fuzzed in the SMS-DELIVER message by related research.

(b) Overview of the fields we fuzzed in the SMS-DELIVER message.

fuzzed in [23]

fuzzed in [24]

fuzzed in [25]

fuzzed in [16,26]

Fig. 3. Overview of the fields fuzzed in the SMS-DELIVER message

The OpenBTS software does not support emergency broadcasts[2], so for these broadcasts we installed a specific branch of an older OpenBTS version (OpenBTS 2.5.4 - SMS-CB), where this service was already implemented.

Having the ability to generate and transmit fuzzed messages, leaves the third stage: the observation. In our SMS fuzzing case, we alternated each fuzzed SMS message with a correct SMS message to see if the phone still responded by acknowledging the correct message. Then after transmitting a batch of alternating

---

[2] Although it is likely that this will be included in a newer release [8].

8	7	6	5	4	3	2	1
Spare	LPD		LB		Sequence Number		
Serial Number (2)							
Message Identifier (2)							
Data Coding Scheme (1)							
Page Parameter (1)							
Content of Message (82)							

Fig. 4. CBS message fuzzing candidates

Table 1. Overview of cell phones tested in this research and the most noticeable results. Legend: I: unremoveable icons, D: DoS message, M: memory bug, N: no notification, R: Reboot S: message handling in violation of specification.

Brand	Type	Firmware/OS	SMS fuzz	Result	CBS fuzz	Result
Apple	iPhone 4	iOS 4.3.3	yes	I,D	no	–
Blackberry	9700	BB OS 5.0.0.743	yes	I	yes	S
HTC	Legend	Android 2.2	yes	I,D	no	–
Nokia	1100	6.64	yes	I	no	–
Nokia	1600	RH-64 v6.90	no	–	yes	S
Nokia	2600	4.42	yes	I,M,R	no	–
Nokia	3310	5.57	yes	I	yes	S
Nokia	3410	5.06	yes	I	no	–
Nokia	6610	4.18	yes	I,N,R	no	–
Nokia	6610	4.74	yes	I,N,R	no	–
Nokia	7650	4.36	yes	I,R	no	–
Nokia	E70-1	3.0633.09.04	yes	I	no	–
Nokia	E71-1	110.07.127	yes	I	no	–
Samsung	SGH-A800	A80XAVK3	yes	I,N,R	no	–
Samsung	SGH-D500	D500CEED2	yes	I,M,R	no	–
Samsung	Galaxy S	Android 2.2.1	yes	I	no	–
Samsung	Galaxy Note	Android 4.1.2	no	–	yes	S
Sony Ericsson	T630	R7A011	yes	I,N	no	–

fuzzed and normal SMS messages we quickly tried most functions of the phones.[3] For our CBS fuzzing we simply used most functions of the tested phones after a batch of fuzzed messages, since there is no acknowledgement of received CBS messages.

Table 1 shows the make and models of the phones we used during the fuzzing research. During the research it turned out that many phones did not support the CBS features, so the test set for CBS fuzzing was small.

4.2 Fuzzing Results

We now give an overview of some of the most interesting results we found during our fuzzing research, which are also summarized in Table 1. For a complete

---

[3] At this stage we also used the phones to empty the SMS memory, which is limited in the older models.

Fig. 5. These two pictures show the strange behavior when the same SMS is opened twice in a row. Note that the words in the left image are the names of games available on the device.

overview of the exact fuzzing performed and the obtained results we refer to the Master's theses on fuzzing SMS [27] and CBS [28] on which this paper is based.

SMS Fuzzing. All tested phones accepted some rarely used SMS variants, such as Fax-over-SMS, which causes strange icons to appear to notify the user of a new message (e.g. a new fax). These SMS variants are so obscure that often the GUI of these mobile phones offered no way for the user to remove these icons, only a message from the network could remove them.

More serious issues were that for five out of sixteen phones we found SMS messages that are received and stored by the phones without any notification to the user. This enables attacks of filling up the SMS memory remotely, but all phones notify the user of a full SMS memory. In addition, seven out of the sixteen tested phones could be forced into a reboot with a single SMS message, though each through a different SMS message.

The Nokia 2600 showed strange behavior where a particular SMS message would display random parts of the phone memory when opened, instead of the SMS message. This behavior is shown in Figure 5.

Both the iPhone 4 and the HTC Legend could be forced in a DoS state were they silently received a message and afterward could no longer receive any SMS messages, without any notification to the user. Rebooting these phones or roaming to a different network would stop the DoS.

Strangely enough we found no real correlation between specific harmful messages and phone brands. So, a message triggering a reboot in a specific Nokia phone, would have no effect on all other Nokia phones. This is likely due to the large variety in phones as explained in Section 3.1

CBS Fuzzing. Our CBS fuzzing research did not reveal any obvious errors such as spontaneous phone reboots. One of the main problems here is that we had no way to tell whether an ignored CBS message was not received by the baseband stack, or that the phone OS did not know how to display it.

The Galaxy Note displayed a fuzzed CBS message which according to the specifications should have been ignored. According to the GSM specification, mobile phones should only receive CBS messages containing Message Identifiers registered in their memory or SIM card. In our initial tests we used the Message Identifier value of 0 and did not register this topic number in the mobile phones. All mobile phones except for the Blackberry received the CBS message. In addition, once we changed the Message Identifier to a value different from 0, all mobile stations did not receive the CBS messages even though this time we did register the topic in the mobile phones.

So we observed that most phones have a lot of trouble to show even correct CBS messages. Since several countries are clearly pushing to get the CBS messages re-supported by phone manufacturers, CBS fuzzing tests should definitely be repeated when wider support is provided.

4.3   Related Work

One of the most well-known bugs found in SMS implementations is the "Curse of Silence" found by Thomas Engel, though it is not directly clear if he used any systematic way, such as fuzzing, to find the vulnerability. With this bug certain Nokia phones stopped receiving SMS messages after receiving an email as SMS message[4] with a sender's email address longer than 32 characters [23].

The most prolific academic researcher in the fuzzing of GSM phones is Collin Mulliner [29,16,26,25]. In 2006 he fuzzed the Multimedia Messaging Service (MMS) feature of GSM [29]. MMS is an extension to SMS for the exchange of multimedia content. When an MMS message is sent the recipient receives an SMS message with a Uniform Resource Identifier (URI) to a server where the MMS content can be retrieved using the Wireless Application Protocol (WAP). Of the three delivery methods discussed op Page 186 Mulliner et al. used the first method by building a virtual (malicious) MMS server using open source software and retrieving content from it on different cell phones. They found several weaknesses in various implementations, including buffer overflows in the Synchronized Multimedia Integration Language (SMIL) parser, the part that takes care of the presentation of the content on the cell phone to the user. Some of these buffer overflows could be used for arbitrary code execution. Mulliner together with Charlie Miller fuzzed SMS messages on smart phones [16,26] using the second method of transmission. The three smart phones available for this research were an iPhone, an Android Phone and a Windows Phone. An application was developed for each of the three platforms, which makes it possible to directly generate and inject SMS messages into the phones modems. Through this application, the researchers were able to make the device believe that an SMS was just received from the GSM network. Finally, Mulliner and

---

[4] Simply this option of receiving email over SMS is a good illustration of how baroque the SMS standard is!

Golde fuzzed the SMS implementation on feature phones [25]. This time they used a rogue cell tower based on OpenBSC, so they used the third method of message transmission. Furthermore they used a J2ME3 application for monitoring on the cell phones. Despite the spectacular title of this publication ("SMS of Death") there was no hard evidence that a fuzzed message caused the death of a phone, since this test was not repeated. The researchers did find DoS attacks for six different popular feature phone brands. They formed SMS messages that can even be sent over commercial (real) networks and will cause the phones to reboot, temporarily losing network connectivity. After consultation with the phone manufacturers Mulliner and Golde did not publish the actual messages that cause the DoS.

The company Codenomicon released a white paper detailing a product that fuzzes SMS messages in order to test the whole network chain for delivery of fuzzed SMS messages [24]. This is targeted towards providers as a tool that can be connected inside the core GSM network.

It is often hard to find out exactly which fields were fuzzed in the studies discussed so far. We have attempted to provide an overview of the fields that have been fuzzed in the SMS-DELIVER message, as far as we can tell from these publications and sometimes through personal communication with the authors. This overview can be found in Figure 3(a). We chose to limit this overview to the one message we also fuzz in our research.

Naturally, fuzzing is not the only way to reveal (security) bugs in the GSM baseband stack. Weinmann et al. decompiled baseband firmware updates from two popular baseband chips and performed a manual code inspection which led to several bugs, amongst which one which led to remote code execution [30,31].

Few researchers have access to the GSM core network. One private security company specializes in fuzz testing GSM core networks and they apparently have a database of possible attacks [32], though the nature of the found vulnerabilities is not public knowledge.

5   Conclusions and Directions for Future Work

We have demonstrated that fuzzing is useful to find bugs in the implementation of GSM stacks on mobile phones. Just think on the number of different mobile phones out in the wild and the information we store on them. Setting up a fake base station and sending out malicious messages is, at least for GSM, not that hard nor expensive anymore and the potential damage could be enormous.

The wide diversity of phones makes it harder to find a single bug affecting many different mobile phones. Nevertheless, our fuzzing research in GSM has shown several issues with mobile phones. The most important attacks here led to various types of DoS messages which can usually be solved through a reboot of the phone. Some results show clear buffer overflow errors, such as the SMS message which will show random parts of the phone's memory when read on the Nokia 2600. Although it is not immediately clear how to abuse such an error for remote code execution, it is possible that such an attack will be constructed in the future for a popular brand of mobile phones. Unfortunately, the CBS service seems to be too poorly supported at the moment to achieve any meaningful fuzzing results, or to use it as an emergency broadcast service for that matter.

The hardest part of fuzzing mobile phone implementations is observing the phone's behavior, which is hard to automate. There are not a lot of other options, other than human testing, for security analysis of the closed source baseband stacks on mobile phones. Then again, with direct access to the baseband stacks fuzz testing these implementations would be much easier. The manufacturers of the baseband stacks or of the SoC have this access and employing strong security tests on their products could greatly increase the security of their product, which among baseband stacks would be a novel selling point. For future research it would be interesting to focus on fuzzing rooted Android devices, where it may be possible to run debuggers in the memory to better observe strange behavior.

For now almost all fuzzing research into GSM has focused on fuzzing mobile phones, and then mostly on fuzzing SMS messages, which still leaves many areas open to explore, such as all the other broadcast messages, but also the network side of a GSM network. It seems logical to assume that the baseband stack on network equipment will contain as many bugs as the stacks on mobile phones, and attacks against the GSM network itself would probably have a much larger impact.

Since the 3G and 4G protocols all have mutual authentication, it is not possible to simply deploy a fake base station in order to fuzz the 3G and 4G baseband stacks. A way around this obstacle would be to use self controlled SIM cards, so you can have your cell tower authenticate to the mobile phone. However, as far as we know there is no open source 3G or 4G cell tower software available yet, so this would require a large amount of work to implement.

Most of our effort came from getting the open source GSM base station up and running. After that implementing the fuzzers was only a few weeks of work. The initial effort to set up a base station and incorporate a fuzzer was substantial, but this solution can now be used to fuzz test any GSM phone on SMS or CBS weaknesses in one and a half hour. This makes fuzzing a cost-effective and feasible technique for making implementations of mobile phone stacks more robust and safe.

References

1. GSM-Association: data and analysis for the mobile industry,
   https://gsmaintelligence.com/
2. UK smartmeter company using GSM/GPRS,
   http://www.smsmetering.co.uk/products/
   smart-meters/gsm-gprs-meters.aspx
3. Hack a day website on sim card carrying traffic lights,
   http://hackaday.com/2011/01/28/sim-card-carrying-traffic-lights/
4. GSM-R Industry Group, http://www.gsm-rail.com/
5. News story on the absence of plans to stop 2g services,
   http://www.computerweekly.com/news/2240160984/
   Will-the-UK-turn-off-its-2G-networks-in-2017

6. Briceno, M., Goldberg, I., Wagner, D.: A pedagogical implementation of the GSM A5/1 and A5/2 "voice privacy" encryption algorithms (1999), http://cryptome.org/gsm-a512.htm (originally on www.scard.org)
7. Website of the Ettus company, selling USRPs, http://www.ettus.com/
8. Burgess, D.: Homepage of the OpenBTS project, http://openbts.sourceforge.net/
9. Nohl, K.: Attacking phone privacy. Blackhat 2010 (2010), https://srlabs.de/blog/wp-content/uploads/2010/07/Attacking.Phone_.Privacy_Karsten.Nohl_1.pdf
10. van den Broek, F., Poll, E.: A comparison of time-memory trade-off attacks on stream ciphers. In: Youssef, A., Nitaj, A., Hassanien, A.E. (eds.) AFRICACRYPT 2013. LNCS, vol. 7918, pp. 406–423. Springer, Heidelberg (2013)
11. ETSI. Digital cellular telecommunications system (Phase 2+); UMTS; LTE; Point-to-Point (PP) Short Message Service (SMS) support on mobile radio interface (3GPP TS 24.011 version 11.1.0 Release 11) (2012)
12. ETSI. Digital cellular telecommunications system (Phase 2+); UMTS;Technical realization of the Short Message Service (SMS), (3GPP TS 23.040 version 11.5.0 Release 11) (2013)
13. Myers, G.J.: The Art of Software Testing. John Wiley & Sons (1979)
14. Kuipers, R., Takanen, A.: Fuzzing embedded devices. GreHack 2012, 38 (2012)
15. Welte, H.: Anatomy of contemporary GSM cellphone hardware (2010), http://laforge.gnumonks.org/papers/gsm_phone-anatomy-latest.pdf
16. Mulliner, C., Miller, C.: Injecting SMS Messages into Smart Phones for Security Analysis. In: Proceedings of the 3rd USENIX Workshop on Offensive Technologies (WOOT). Montreal, Canada (August 2009)
17. van den Broek, F., Wichers Schreur, R.: Femtocell Security in Theory and Practice. In: Riis Nielson, H., Gollmann, D. (eds.) NordSec 2013. LNCS, vol. 8208, pp. 183–198. Springer, Heidelberg (2013)
18. Welte, H.: Homepage of the OpenBSC project, http://openbsc.osmocom.org/
19. Homepage of the GNU Radio project, http://gnuradio.org/
20. Welte, H.: Homepage of the OsmoBTS project, http://openbsc.osmocom.org/trac/wiki/OsmoBTS
21. Collection of fuzzing software, http://fuzzing.org/
22. Code archive of the sulley fuzzing framework, https://github.com/OpenRCE/sulley
23. Engel, T.: S60 Curse of Silence. CCC Berlin (2008) http://berlin.ccc.de/~tobias/cos/
24. Vuontisjärvi, M., Rontti, T.: SMS Fuzzing. Codenomicon whitepaper (2011), http://www.codenomicon.com/resources/whitepapers/codenomicon_wp_SMS_fuzzing_02_08_2011.pdf
25. Mulliner, C., Golde, N., Seifert, J.-P.: SMS of Death: From Analyzing to Attacking Mobile Phones on a Large Scale. In: USENIX (2011)
26. Mulliner, C., Miller, C.: Fuzzing the Phone in your Phone. Black Hat USA (June 2009)
27. Hond, B.: Fuzzing the GSM protocol. Master's thesis, Radboud University Nijmegen, Kerckhoff's Master, The Netherlands (2011)
28. Torres, A.C.: GSM cell broadcast service security analysis. Master's thesis, Technical University Eindhoven, Kerckhoff's Master, The Netherlands (2013)

29. Mulliner, C., Vigna, G.: Vulnerability Analysis of MMS User Agents. In: Proceedings of the Annual Computer Security Applications Conference (ACSAC), Miami, FL (December 2006)

30. Weinmann, R.-P.: Baseband Attacks: Remote Exploitation of Memory Corruptions in Cellular Protocol Stacks. In: WOOT, pp. 12–21 (2012)

31. Weinmann, R.-P.: The baseband apocalypse. In: 27th Chaos Communication Congress Berlin (2010)

32. P1Security. website detailing a fuzzing product for telco core-networks, http://www.p1sec.com/corp/products/p1-telecom-fuzzer-ptf/

User-Centric Security Assessment of Software Configurations: A Case Study

Hamza Ghani, Jesus Luna Garcia, Ivaylo Petkov, and Neeraj Suri

Technische Universität Darmstadt, Germany
{ghani,jluna,petkov,suri@deeds.informatik.tu-darmstadt.de}

Abstract. Software systems are invariably vulnerable to exploits, thus the need to assess their security in order to quantify the associated risk their usage entails. However, existing vulnerability assessment approaches e.g., vulnerability analyzers, have two major constraints: *(a)* they need the system to be already deployed to perform the analysis and, *(b)* they do not consider the criticality of the system within the business processes of the organization. As a result, many users, in particular small and medium-sized enterprizes are often unaware about assessing the actual *technical and economical impact* of vulnerability exploits in their own organizations, *before* the actual system's deployment. Drawing upon threat modeling techniques (i.e., attack trees), we propose a user-centric methodology to quantitatively perform a software configuration's security assessment based on *(i)* the expected economic impact associated with compromising the system's security goals and, *(ii)* a method to rank available configurations with respect to security. This paper demonstrates the feasibility and usefulness of our approach in a real-world case study based on the Amazon EC2 service. Over 2000 publicly available Amazon Machine Images are analyzed and ranked with respect to a specific business profile, before deployment in the Amazon's Cloud.

Keywords: Cloud Security, Economics of Security, Security Metrics, Security Quantification, Vulnerability Assessment.

1 Introduction

The use of information systems has been proliferating along with rapid development of the underlying software elements driving them (e.g., operating systems and commercial off-the-shelf software). However, this rapid development comes at a cost, and in many cases e.g., due to limited time schedules and testing budgets for releasing new products, software is often not rigorously tested with respect to security. This results in security flaws that can be exploited to compromise the confidentiality (C), integrity (I) and availability (A) of the affected software products. These flaws are referred to as *software vulnerabilities* and are collected, quantitatively scored and categorized by a multitude of vulnerability databases (e.g., the National Vulnerability Database NVD [1] or the Open Source Vulnerability Database OSVDB [2]). It is a prevalent practice to assess

J. Jürjens, F. Piessens, and N. Bielova (Eds.): ESSoS 2014, LNCS 8364, pp. 196–212, 2014.
© Springer International Publishing Switzerland 2014

the security of a software system using software analyzers (e.g., OpenVAS [3] and Nessus [4]), that query databases like NVD to ascertain the vulnerabilities affecting a specific software configuration (cf., Figure 1). Unfortunately, despite their broad usage, this approach has two main drawbacks:

1. Most (if not all) vulnerability analyzers require the *deployed* software system to perform the assessment, therefore resulting in a costly trial-and-error process.
2. Such security assessment does not take into account the *economic impact* of detected vulnerabilities. Therefore, it is common to find inconsistencies e.g., technically critical vulnerabilities that do not have the highest economic impact on the organization [5].

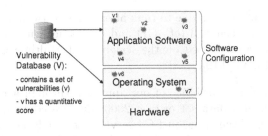

Fig. 1. System Model - Software Configurations and Vulnerabilities (v_i)

Empirical research has shown that the actual impact of vulnerability exploits varies significantly among different types of organizations (in particular the smaller/medium enterprises or SMEs) [6,7]. Since different organizations perceive the severity of a particular vulnerability differently, they also prioritize its mitigation differently. Existing hypotheses advocate user-centric approaches [8], where the quality and customization of the performed security assessments can be improved if these correlate to the user awareness on the actual impact of a vulnerability in their particular organizational context.

In order to empower users to perform an accurate assessment and ranking of available software configurations *before deploying them*, we propose a methodology to perform the security assessment of a software configuration based on the user's organizational context (expressed in the form of both expected technical and economical impacts). Figure 2 depicts the main stages of our proposed approach, where the specific paper contributions are:

– C1: An approach to elicit the technical metrics required to quantitatively reason about the security goals (C, I, A) of a software configuration, based on the notion of threat modeling and attack trees.
– C2: A systematic approach eliciting the economic-driven factors for weighting the user's security goals, in order to improve the conclusions that can be drawn from the generated attack trees (cf., C1).

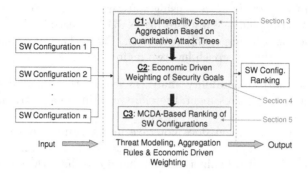

Fig. 2. Overview of the Proposed Approach

- C3: A quantitative technique to rank alternative software configurations using as input the technical (cf., C1) and economical metrics (cf., C2). Our ranking technique is based on the widely used Multiple Criteria Decision Analysis (MCDA) [9,10].

We demonstrate the feasibility of our approach through a real-world Cloud case study, in which a data set of over 2000 software configurations (publicly available for users of the Amazon EC2 service) are analyzed and ranked from a security perspective before the actual deployment. The contributed approach aims to enhance the usefulness of widely used security analyzers, by providing users with additional tools that take into account their own organizational contexts.

The remainder of this paper is organized as follows: Section 2 introduces a motivating case study. Sections 3 – 5 detail the paper contributions as depicted in Figure 2. The results of our evaluation using real world data from Amazon EC2 are shown in Section 6. Section 7 summarizes existing related approaches and Section 8 provides conclusions for the paper.

2   Motivating Case Study: Security-Aware Selection of Amazon Machine Images

While the many economic and technological advantages of Cloud computing are apparent, the migration of key business applications onto it has been limited, in part, due to the lack of security assurance on the Cloud Service Provider (CSP). For instance the so-called Infrastructure-as-a-Service (IaaS) Cloud providers allow users to create and *share* virtual images with other users. This is the case of e.g., Amazon EC2 service where users are given the chance to create, instantiate, use and share an Amazon Machine Image (AMI) without the hassle of installing new software themselves. The typical IaaS trust model considers that users trust the CSP (e.g., Amazon), but the trust relationship between the provider of the virtual image –not necessarily Amazon – and the user is not as clear [11].

The basic usage scenario for the Amazon EC2 service requires the user to access the "AWS Management Console" in order to search, select and instantiate the AMI that fulfills her functional requirements (e.g., specific software configuration, price, etc.). Even in simple setups, the security of the chosen AMI (e.g., number and criticality of existing vulnerabilities) remains unknown to the customer *before* its instantiation. Once instantiated, it is the responsibility of the user to assess the security of the running AMI and take the required measures to protect it. However, in a recent paper Balduzzi [11] demonstrated that both the users and CSPs of public AMIs may be exposed to software vulnerabilities that might result in unauthorized accesses, malware infections, and loss of sensitive information. These security issues raise important questions e.g., is it possible for an Amazon EC2's user to assess the security of an AMI before actually instantiating it? Or, can we provide an Amazon EC2's customer with the AMI that both fulfills the functional requirements and, also represents the smallest security risk for the organization?

3 Vulnerability Score Aggregation Based on Attack Trees

This section presents the first contribution of the proposed assessment methodology (cf., Stage 1 in Figure 2), as an approach to quantify the *aggregated* impact of a set of vulnerabilities associated with a software configuration, based on the notion of *attack trees* [12]. Quantified technical impact and proposed economic metrics (cf. Section 4), will be used as inputs to the MCDA methodology (cf. Section 5) to rank available software configurations.

3.1 Building the Base Attack Pattern

Taking into account that the basic concepts of threat modeling are both well-documented (see Section 7 for more details) and broadly adopted by the industry (e.g., Microsoft's STRIDE threat modeling methodology [13]), the initial stage of the proposed methodology is built utilizing the notion of attack trees. Attack trees, as also used in our paper, are hierarchical representations built by creating nodes that represent the *threats* to the software configuration i.e., the security properties that the attacker seeks to compromise (any of C, I or A). Then one continues adding the *attack* nodes, which are the attacker's strategies to pose a threat to the system (e.g., Denial of Service, SQL injection, etc.). Finally, the attack tree's leaf nodes are populated with the actual software *vulnerabilities* that might be exploited by the attacker to launch an attack. As mentioned in Section 1, software vulnerabilities are associated with a unique identifier and a numeric score similar to those in contemporary databases e.g., NVD [1] and OSVDB [2].

One of the main advantages related with the use of attack trees, is that they allow the creation of "attack tree patterns". The usefulness of attack tree patterns has been documented by the U.S. Department of Homeland Security [14] and, also has been researched in EU projects e.g., SHIELDS [15]. The attack

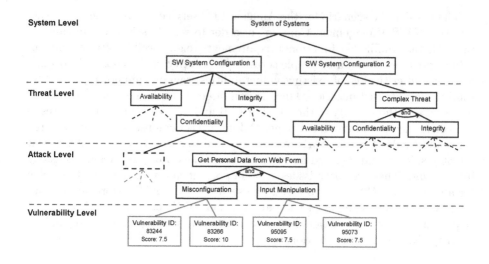

Fig. 3. Extended attack tree pattern

tree built with the basic information presented in this section will be called *base attack pattern* in this paper. Section 6 will introduce a tool we have developed to automatically create attack trees based on the output of the Linux RPM package manager [16].

Base attack patterns can be re-used or even extended by other users to model their own organizational contexts/concerns, therefore taking advantage of the knowledge from the experts that originally created them. For example, our base attack pattern can be further extended with the different elements shown in Figure 3 (i.e., AND nodes[1], composite attacks/threats). The conclusions that can be drawn from the attack tree shown in Figure 3 (i.e., the aggregated impact of a software configuration's vulnerabilities), can be greatly improved if we provide the techniques to quantitatively reason about the numeric scores associated with each node, as presented in the next section.

3.2   Quantitatively Reasoning about Attack Trees

The proposed rules for aggregating the numeric scores in an "extended attack tree pattern" (cf., Figure 3), requires that every software vulnerability in the tree has a score (similar to NVD [1]). If the vulnerability does not currently have a score, then predictive techniques like VAM [17] can be utilized to propose or predict a value. Based on widely used scoring systems like CVSS [18], we also make the conventional assumption that the provided vulnerability scores are on the interval [0, 10]. The aggregation rules proposed in this paper (cf., Table 1) are

---

[1] The AND relationship is only an example option and more complex logical rules can be set up by the user as needed for their applications context.

Table 1. Aggregation Rules for Attack Trees

Relationship	Aggregation rule for node N
AND	$Agg_N = \frac{\sum_{i=1}^{m} N_i}{m}$ where $m = N$'s number of children nodes
OR	$Agg_N = max(N_1 \ldots N_m) \times \frac{m}{n}$ where $m = N$'s number of children nodes n = total number of nodes at the same level than N's children $(n \geq m)$

recursively applied throughout the attack tree in a bottom-up approach, starting at the vulnerability level and finishing at the threat level (cf., Figure 3). Our proposed aggregation rules are based on previous research in the Privacy-by-Design [19] and Cloud security metrics topics [20], and only need to differentiate the actual relationship among the siblings (i.e., AND/OR). Future work will analyze the effect of aggregating at a higher level of granularity on the attack tree (i.e., the system level). A detailed example on the use of extended attack tree patterns and, designed aggregation rules is presented in Section 6.

4 Economic Driven Weighting of Security Goals

In this section we investigate the trade-offs between security and those economic considerations that play a central role in the proposed methodology.

4.1 Including The Economic Perspective

As information systems constitute a mean for helping organizations meet their business objectives, not considering economic aspects when assessing IT security is potentially a major issue (e.g., reputation loss caused by vulnerability exploits). As required by our model, in order to determine the user priorities w.r.t. security goals (i.e., C, I, A) and their relative importance, we propose to use a novel economic driven approach. The rationale is that the potential economic damage to the business caused by a security compromise determines significantly the weight of the security goals. As security goals do not equally influence the core business of the considered organization, they need to be quantitatively weighted following a user-centric approach taking into account the business context specificities. Next, we elaborate on the economic driven damage estimation metrics suitable for weighting an organization's security goals.

4.2 Running Example – Business Profiling

In this section we introduce a calculation model for weighting the security goals based on the notion of "business profile", which refers to the organization's *(i)*

Table 2. Business Profiles: Qualitative/Quantitative Assessment

Characteristics	SME X		Multinational Y	
	Qualitative	Quantitative	Qualit.	Quant.
Organization Size (OS)	30	$\frac{1}{3}$	5500	1
Sector of Activity (SA)	Manufacturing	$\frac{1}{3}$	Direct Banking	1
Countries of Activity (CA)	Mexico	$\frac{1}{3}$	US, Euro zone	1
Security Policy (SP)	No	$\frac{1}{3}$	E & M	1
Annual Turnover (AT)	3M USD	$\frac{1}{3}$	750M USD	1
Customer Data (CD)	Personal Data	$\frac{2}{3}$	P & F	1
Employee Data (ED)	P & F	1	P & F	1
Intellectual Property (IP)	No	$\frac{1}{3}$	Risk Models	$\frac{2}{3}$

economic and *(ii)* data-centric characteristics (as suggested by the authors of [21]). Both set of characteristics, altogether denoted as CH, are the basis for evaluating the weights for the cost categories depicted in Figure 4 and Table 2. In analogy to widely used scoring systems like CVSS [18] and taking into account related works [21], we propose the following eight CH and the corresponding set of qualitative values:

- $OS = \{$Less than $50 <$ Less than $250 <$ More than 250 employees$\}$
- $SA = \{$Low $<$ Moderate $<$ High IT dependency$\}$
- $CA = \{$Others $<$ Euro zone $<$ United States$\}$
- $SP = \{$No $<$ Existing $<$ Existing & Monitored (E & M)$\}$
- $AT = \{$less $10M <$ less $50M <$ more $50M$ USD$\}$
- $CD = \{$No $<$ Personal Data $<$ Personal & Financial (P & F)$\}$
- $ED = \{$No $<$ Personal Data $<$ Personal & Financial$\}$
- IP (patents, blueprints, etc.) $= \{$No $<$ Moderate value $<$ High value$\}$

To perform the calculation process these qualitative values in CH will be mapped to quantitative values (e.g., $(\frac{1}{3}; \frac{2}{3}; 1)$ as used in this section). To illustrate our approach, let us consider an example with two companies i.e., *(i)* an SME X, and *(ii)* a large multinational company Y. Both have their respective company profiles depicted in Table 2. Thanks to the proposed approach, whatever company C can be represented as a tuple $C = (OS, SA, CA, SP, AT, CD, ED, IP)$ containing the quantitative values of the characteristics of C. For instance, the SME X shown in Table 2 can be represented by the tuple $(\frac{1}{3}, \frac{1}{3}, \frac{1}{3}, \frac{1}{3}, \frac{1}{3}, \frac{2}{3}, 1, \frac{1}{3})$. The notion of business profiles will be utilized as a basis to weight the economic driven metrics to be defined in the next section.

4.3 Economic Driven Approach for Weighting Security Goals

The methodology proposed in this paper requires a set of metrics reflecting the economic impact of potential security incidents, caused by software vulnerability exploits. To define these *Economic Driven Metrics* (EDM), one needs to

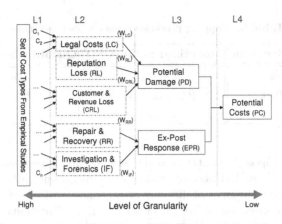

Fig. 4. Aggregation of the Proposed Economic Driven Metrics

investigate the expected potential costs of security incidents. The main basis for determining our set of applicable EDM is the work of Innerhofer et al. [22], in which the authors define a set of 91 cost units based on an empirical study on the costs caused by publicly known security incidents. Because in many cases (security) managers are in charge of assessing the economic impact of vulnerabilities, experience has proved that it is more convenient to evaluate higher/aggregated levels of granularity for the potential costs of a security exploit in order to be intuitive and easy to classify. Therefore, we define a small set of higher level main cost classes aggregating the 91 cost units of [22,23]. We distinguish three levels of granularity regarding the potential costs: the highest level (most detailed) is $L1$ which contains all cost units defined in the state of the art literature [22,23]. $L2$ aggregates the $L1$ costs into one of the five proposed cost classes (dashed rectangles in Figure 4) with context-dependent weights $((W_{LC}), (W_{RL})$, etc.) reflecting the criticality of the corresponding cost (sub)class for the organization. For instance "Legal Costs (LC)" are highly dependent on the country, in which the organization is located, thus need to be weighted differently in diverse legal environments (e.g., the jurisprudence in the USA is completely different than in Germany, developing countries, etc.). The lowest level of granularity $L3$ distinguishes between two main cost classes (i) potential damage/losses, and (ii) ex-post response costs, which could result from a security incident. Figure 4 depicts the overall cost aggregation process. In earlier work, we have described our methodology for systematically investigating all cost units and the corresponding unified cost classes [24], which constitute the underlying cost data for $L1$ (Figure 4).

Based on the same principles that CVSS [25], we propose to use an intuitive scale of three possible values (i.e., low, medium, high) to evaluate the different metrics of level $L2$ (cf. Figure 4). Furthermore, as monetized metrics have the advantage of (i) allowing easy numerical comparison between alternative scenarios within the same company, and (ii) are directly understandable by managers and executives with less technical affinity, we propose a mapping (cf., Table

Table 3. Mapping proposed scale - monetized scale

Qualitative Scale	Monetized Scale (USD)	Quantitative Scale
Low	$[0, C_{medium}[$	3.5
Medium	$[C_{medium}, C_{high}[$	6.1
High	$[C_{high}, \infty[$	9.3

3) between our proposed qualitative scale and a company-dependent monetized scale. The rationale is that absolute monetary terms do not allow an objective comparison across companies of different sizes; e.g., a cost of $100K$ EUR might be *critical* for an SME, but of *low* effect for a large multinational company.

Organizations could define their specific interval values c_x for the monetized mapping. For the calculation of our metrics, one needs also quantified factors to be mapped to the proposed scale (cf., Table 3). The quantitative scale thresholds are defined in such a way that, analogous to the CVSS thresholds, the scoring diversity is taken into consideration [26] and the intuitive and widely accepted CVSS scoring scheme is respected. To illustrate the usage of the metrics e.g., for "Reputation Loss" in the case of a *Confidentiality* compromise, the user can qualitatively estimate a value (i.e., low, medium, high) for that specific EDM, and according to the mapping depicted in Table 3, a quantitative value to be utilized for the score calculations is assigned accordingly. To calculate the metric for the overall "Potential Costs" ($L4$) we define the final outcome of calculating PC for *each* security goal (i.e., C, I, A) as depicted in Equation 1 for Confidentiality (similarly for I, A):

$$PC_C = (LC \times W_{LC}) + (RL \times W_{RL}) + (CRL \times W_{CRL}) + (RR \times W_{RR}) + (IF \times W_{IF}) \quad (1)$$

Furthermore, the different weights for $L2$ (Figure 4) needed to compute Equation 1 are calculated as follows, where $Max(X)$ is the maximal possible value for X:

$$W_{LC} = \frac{CA}{Max(CA)} \quad (2)$$

$$W_{RL} = \frac{OS + SA + CA + AT}{Max(OS) + Max(SA) + Max(CA) + Max(AT)} \quad (3)$$

$$W_{CRL} = \frac{AT + CD + IP}{Max(AT) + Max(CD) + Max(IP)} \quad (4)$$

$$W_{RR} = \frac{OS + SP}{Max(OS) + Max(SP)} \quad (5)$$

$$W_{IF} = \frac{OS + SP + CD + ED + IP}{Max(OS) + Max(SP) + Max(CD) + Max(ED) + Max(IP)} \quad (6)$$

To calculate these weights, we utilize the business profiling values defined in Table 2. The weight calculations for SME X and Multinational Y provide the results shown in Table 4. In the next section, we introduce the last stage of our approach consisting of an MCDA-based approach to assess and rank different software configurations, taking as input the outcomes of Sections 3 and 4.

Table 4. $L2$ Weights for SME X and Multinational Y

Weights ($L2$)	SME X	Multinational Y
W_{LC}	$\frac{1}{3}$	1
W_{RL}	$\frac{1}{3}$	1
W_{CRL}	$\frac{4}{9}$	$\frac{8}{9}$
W_{RR}	$\frac{1}{3}$	1
W_{IF}	$\frac{8}{15}$	$\frac{14}{15}$

5   MCDA-Based Ranking of Software Configurations

In this section, we present a MCDA-based methodology by which the proposed raking of software configurations can be performed in a systematic way. MCDA methods are concerned with the task of ranking a finite number of alternatives (software configurations in our case), each of which is explicitly described in terms of different characteristics (i.e., the aggregated vulnerability scores from Section 3) and weights (i.e., the economic-driven metrics from Section 4) which have to be taken into account simultaneously. For our research we decided to apply the Multiplicative Analytic Hierarchy Process (MAHP) [9,27], one of the most widely used and accurate MCDA methodologies nowadays [10]. In the following, the MAHP-background required to comprehend our approach will be briefly presented . For a detailed description of the MCDA methods (including MAHP), we refer to [28].

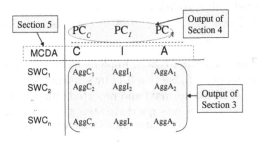

Fig. 5. MAPH-based matrix used by our approach

At a glance, MAHP starts by building a matrix as shown in Figure 5 in order to perform the ranking. The MAHP matrix requires the aggregated impacts of a set of vulnerabilities associated with a software configuration. Furthermore, the EDMs that have been introduced in Section 4 constitute the weights (PC) of the security goals to take into account (i.e., C, I, A). Once the MAHP matrix is built, we calculate a quantitative score S_{SWC_i} for each software configuration utilizing Equation 7. The value of S_{SWC_i} is directly proportional to the overall impact (technical and economical) associated with the software configuration i.e., a low S_{SWC_i} represents also a low impact for the organization. In the next section, we will experimentally show how thanks to MAHP it is possible to quantitatively rank different Amazon EC2's AMIs configurations from a user-centric perspective.

$$S_{SWC_i} = (AggC_i)^{PC_C} \times (AggI_i)^{PC_I} \times (AggA_i)^{PC_A} \tag{7}$$

6   Evaluation: Security Ranking of Amazon EC2's AMIs

Further developing the Amazon EC2-based case study introduced in Section 2, performed validation experiments and obtained results are presented next.

6.1   Experimental Setup

Our validation experiments consider a SME user of the Amazon EC2 service (cf., Section 2), who is looking for an available AMI with a LAMP software configuration[2]. Our methodology aims to provide this user with quantitative security insights about alternative AMIs *before instantiating any*. In particular we will take into consideration for the assessment her organizational context (i.e., technical and economical risks).

The proposed methodology was validated using real-world vulnerability data (i.e., Nessus' reports [4]) from more than 2000 Amazon EC2's AMIs, kindly provided for research purposes (i.e., sanitized and anonymized) by Balduzzi et. al. [11]. It is also worth to highlight that this data set covers a period of five months, between November 2010 and May 2011, and as mentioned by Balduzzi [11] the Amazon Web Services Security Team already took the appropriate actions to mitigate the detected vulnerabilities.

The implemented test bed (cf., Figure 6) consisted of three main building blocks, namely:

- The "Attack Tree", a Java/MySQL implementation to automatically create "base attack patterns" (cf., Section 3.1) by sequentially extracting both AMI configurations (RPM-like format) and reported vulnerabilities from the data set (Step 1a). OSVDB [2] was queried (Step 1b) to classify found vulnerabilities into corresponding attacks, and then NVD [1] scores were used to compute our "coverage" metric (cf., Section 6.2). Finally, this component also aggregated the values on the resulting attack tree applying the rules presented in Section 3.2.
- The "Economic Metrics" component (web form and back-end database) where the User inputs the information related to her own organizational context (Step 2). This information is processed to create the numeric weights (i.e., PC_C, PC_I and PC_A presented in Section 4.3) required by the ranking module described next.

---

[2] LAMP stands for the software system consisting of Linux (operating system), Apache HTTP Server, MySQL (database software), and PHP, Perl or Python.

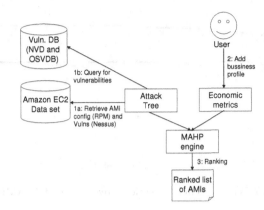

Fig. 6. Test bed used for validation

- The "MAHP engine" implements the MAHP technique described in Section 5, which takes as inputs both the aggregated technical impact (from the "Attack Tree" component) and the economic-driven weights (from the "Economic Metrics" module). The output is an ordered set of AMIs.

For our ranking experiments, we used the two synthetic business profiles shown in Table 2 (i.e., SME X and Multinational Y). At the time of writing this paper we still do not have the information for creating real-world profiles, however as discussed in Section 8 we have started collecting this data via targeted surveys.

6.2   Evaluating the Methodology's Coverage

The goal of this experiment was to validate if the vulnerabilities reported by our approach (cf., Step 1b in Figure 6) were *at least* the same as reported by the Nessus tool. If that is the case, then we can actually assert the validity of performing the proposed AMI's security assessment *before* instantiation. Obtained coverage results are shown in Figure 7 for all tested 2081 AMIs. A coverage rate of at least 90% was achieved in 93.46% of the AMIs, with a worst case scenario of 65% coverage in only *one* AMI.

One of the main challenging issues facing our current implementation is ensuring 100% coverage. Vulnerabilities are queried from publicly available databases (e.g., NVD [1]) based on a mapping between the actually installed software (RPM-like format [16]) and, the Nessus reported software (using the Common Platform Enumeration or CPE format [29]). Therefore, we cannot claim that our mapping is complete, as it does not contain all existing software packages. Unfortunately, at the state of the art there is no publicly available $RPM \leftrightarrow CPE$ mapping that can be applied for this purpose. So we had to manually check and complement the mapping to run our experiments meaningfully. Such a comprehensive/constantly updated mapping, could allow Amazon EC2 to actually provide its users with a realistic security assessment of existing AMIs (before instantiation).

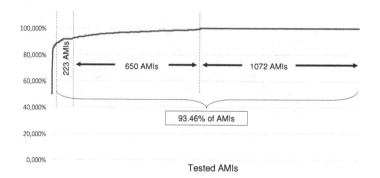

Fig. 7. Vulnerability Detection Coverage in 2081 AMIs

6.3   Ranking Existing AMIs

During this experiment, we applied the business profiles presented in Table 2 to the data set of 2081 AMIs in order to rank them with the MAHP technique described in Section 5. As required by the MAHP matrix (cf., Figure 5), the base attack pattern for each available AMI was automatically created and populated in order to compute the aggregated impacts Agg_{Ci}, Agg_{Ii} and Agg_{Ai}. For the sake of automation, our experiments did not extend the base attack pattern (e.g., with the use of AND relationships).

Just as expected, ranking results show that for both business profiles (i.e., SME X and Multinational Y) the order of the best suitable AMIs is different. For instance the 2nd best AMI for SME X is ranked 20 for Multinational Y, and the 3rd one for SME is ranked 21st for the latter. Table 5 depicts the rankings of the top 10 AMIs in both scenarios (SME X and Multinational Y). Notice that in both scenarios the best ranked AMI was the same (`ami-fb6e8292`), because this configuration has both the least number of critical vulnerabilities and, relatively

Table 5. Top 10 AMI Rankings: SME X vs. Multinational Y

Rank	Multinational Y	MAHP Score	SME X	MAHP Score
1	*ami-fb6e8292*	0.231468538	*ami-fb6e8292*	0.738087398
2	ami-f857b091	0.3934929	ami-044fa56d	1.002042157
3	ami-43aa432a	0.534172853	ami-2309e44a	1.002042157
4	ami-63aa430a	0.534172853	ami-49c72920	1.002042157
5	ami-665bb00f	0.534172853	ami-6c749e05	1.002042157
6	ami-6743ae0e	0.534172853	ami-8f729fe6	1.002042157
7	ami-7d43ae14	0.534172853	ami-a236c1cb	1.002042157
8	ami-8ff38cdd	0.534172853	ami-43aa432a	1.014506003
9	ami-bb709dd2	0.534172853	ami-63aa430a	1.014506003
10	ami-c224d5ab	0.534172853	ami-665bb00f	1.014506003

low aggregated scores in the corresponding attack tree. The latter also explains the low overall score S_{SWC_i} obtained by the MAHP technique (despite the two different organizational profiles). This quantitative result proves the intuitive notion that a properly secured AMI, can provide an adequate security level to different types of users/organizations.

7 Related Work

Despite the large variety of papers devoted to vulnerability assessment in software systems, there are, to the best of our knowledge, no existing approaches that take account of both technical and economical perspectives in the assessment process. Nevertheless, we present in this section relevant existing research in the field of vulnerability assessment. For the vulnerability assessment approaches from a technical perspective, there exist varied academic and applied approaches to manage and assess vulnerabilities. Projects described in [30,31,32] define a list of detected vulnerabilities, typically ranked using qualitative assessment such as low, medium, high. These assessment approaches have a qualitative nature and do not consider economic aspects. Mell et. al. [33] propose quantitative metrics for calculating scores reflecting the severity of vulnerabilities based on published vulnerability reports. They introduce CVSS, which is a vulnerability scoring system providing a standardized method for rating vulnerabilities [26]. Our approach can be used to add the necessary contextual dimension to improve the usage and accuracy of CVSS scores (cf., Section 3). This aspect of our approach is important, when considering the existing works suggesting that different organizations evaluate vulnerabilities differently, based on their specific contexts [34,8,35,36]. The authors of [6,7] showed empirically that the impact of security vulnerability exploits varies with a company's context. The results of [6,7] constitute a major driver motivating our work.

Like our approach, there exists a separate line of research applying MCDA techniques in security related fields. The authors of [37] utilize MAHP for the security assessment of power control processes. Similarly, another work in the area of power systems [38] applies an MCDA approach to provide online quantification of a security level associated with an existing or forecasted operating condition of power systems.

In the area of economic driven metrics, the authors of [39] analyzed the incurred costs related to the resolution of individual security incidents within 18 participating US schools. Additionally, there are simple calculators of potential losses such as "Data Breach Risk Calculator" [21] from the Ponemon Institute and Symantec Corporation, and the "Data Loss Cost Calculator" [40]. These calculators provide rough numbers (mostly for illustrative purposes), and their calculation formulas and methodologies are mostly hidden. Another related field concerns cybercrime and its economic impact on the society. Anderson [41] introduces the first systematic study of the costs of cybercrime in the society, as an answer to the the UK Detica report [42]. Clearly, our approach is hence the first vulnerability assessment method that uses both technical and economic driven metrics in the calculation process and aggregates them in a holistic manner, enabling a user centric, pre-deployment assessment of security vulnerabilities of different software configurations.

8 Conclusions

In this paper we presented a methodology to quantitatively assess the security of a software configuration from a user-centric perspective. The proposed approach takes into account the overall organizational context (i.e., technical and economical risks), and does not require the actual software system to be deployed/installed. The proposed approach has been validated using real-world data from Amazon EC2. Vulnerability reports covering a total number of 2081 AMIs has been considered in the evaluation of our approach. The obtained results show that (i) our approach *not requiring a physical system deployment* is able to report *at least* the same vulnerabilities as Nessus [4] (a coverage rate of 93.46% of the tested AMIs); and (ii) given some business profiling data (e.g., turn-over, countries of activity), it is feasible to rank available Amazon EC2's AMIs with respect to security. While a limitation of our approach consists in that the results' accuracy depends on the quality of the available input data, especially business profiling data, our findings suggest that the proposed assessment could be adopted with little effort by IaaS providers, thus empowering their customers to compare different existing configurations and offers from a security level perspective.

Furthermore, we are investigating alternative aggregation rules to be applied on the attack trees, a promising direction consists in utilizing semi-ring operations in order to interpret the "AND" branches.

Acknowledgments. Research supported in part by EC FP7 SPECS, Loewe-CASED and BMBF EC-SPRIDE at TU Darmstadt. The authors would like to thank Marco Balduzzi, Jonas Zaddach, and especially Davide Balzarotti, Engin Kirda and Sergio Loureiro for sharing with us the Amazon data set for our experiments.

References

1. NVD, National Vulnerability Database (2013), http://nvd.nist.gov/
2. OSVDB, The Open Source Vulnerability Database (2012), http://osvdb.org/
3. OpenVAS, Open Vulnerability Assessment System (2013), http://www.openvas.org/
4. Tenable Network Security, Nessus vulnerability scanner (2013), http://www.tenable.com/products/nessus
5. Fruehwirth, C., et al.: Improving CVSS-based vulnerability prioritization and response with context. In: Proc. of Third International Symposium on Empirical Software Engineering and Measurement (2009)

6. Ishiguro, M., et al.: The effect of information security incidents on corporate values in the japanese stock market. In: Proc. of International Workshop on the Economics of Securing the Information Infrastructure, WESII (2006)

7. Telang, R., et al.: An empirical analysis of the impact of software vulnerability announcements on firm stock price. Proc. of IEEE Transactions on Software Engineering (2007)

8. Lai, Y., et al.: Using the vulnerability information of computer systems to improve the network security. Computer Communications (2007)

9. Saaty, T.: Book: The Analytic Hierarchy Process. McGraw-Hill, New York (1980)

10. Triantaphyllou, E.: The impact of aggregating benefit and cost criteria in four mcda methods. IEEE Transactions on Engineering Management (2004)

11. Balduzzi, M., et al.: A security analysis of Amazon's Elastic Compute Cloud service. In: Proc. of the Annual ACM Symposium on Applied Computing (2012)

12. Schneier, B.: Attack trees. Dr Dobb's 24(12) (1999),
http://www.schneier.com/paper-attacktrees-ddj-ft.html

13. Swiderski, F., Snyder, W.: Book: Threat Modeling. Microsoft Press (2004)

14. Department of Homeland Security, Attack Patterns (2009),
https://buildsecurityin.us-cert.gov/

15. SHIELDS, EU FP 7 – SHIELDS project: Detecting known security vulnerabilities from within design and development tools (2010),
http://www.shields-project.eu/

16. RPM ORG, The RPM package manager (2007), http://rpm.org/

17. Ghani, H., et al.: Predictive vulnerability scoring in the context of insufficient information availability. In: Proc. of the Intl. Conference on Risk and Security of Internet and Systems, CRiSIS (2013)

18. Forum of Incident Response and Security Teams, CVSS – Common Vulnerability Scoring System (2012), http://www.first.org/cvss/

19. Luna, J., et al.: Privacy-by-design based on quantitative threat modeling. In: Proc. of the Intl. Conference on Risk and Security of Internet and Systems (2012)

20. Luna, J., et al.: Benchmarking Cloud Security Level Agreements Using Quantitative Policy Trees. In: Proc. of the ACM Cloud Computing Security Workshop (2012)

21. Symantec, Ponemon Institute, Data Breach Calculator (2013),
https://databreachcalculator.com

22. Innerhofer, F., et al.: An empirically derived loss taxonomy based on publicly known security incidents. In: Proc. of Intl. Conf. on Availability, Reliability and Security, ARES (2009)

23. Van Eeten, M., et al.: Damages from internet security incidents. OPTA Research reports (2009),
http://www.opta.nl/nl/actueel/alle-publicaties/publicatie/?id=3083

24. Ghani, H., et al.: Quantitative assessment of software vulnerabilities based on economic-driven security metrics. In: Proc. of the Intl. Conference on Risk and Security of Internet and Systems, CRiSIS (2013)

25. Forum of Incident Response and Security Teams, CVSS Adopters (2013),
http://www.first.org/cvss/eadopters.html.

26. Scarfone, K., Mell, P.: An analysis of CVSS version 2 vulnerability scoring. In: Intl. Symposium on Empirical Software Engineering and Measurement, ESEM (2009)

27. Saaty, T.: Book: Fundamentals of decision making and priority theory with the analytic hierarchy process. RWS Publications, Pittsburgh (1994)

28. Zeleny, M.: Book: Multiple Criteria Decision Making. McGraw-Hill (1982)

29. NIST, CPE – Official Common Platform Enumeration Dictionary (2013),
    http://nvd.nist.gov/cpe.cfm
30. SANS-Institute, SANS critical vulnerability analysis archive (2007),
    http://www.sans.org/newsletters/cva/
31. Johnson, E., et al.: Symantec global internet security threat report (2008),
    http://eval.symantec.com/mktginfo/enterprise/white_papers/
    b-whitepaper_internet_security_threat_report_xiii_04-2008.en-us.pdf
32. Microsoft, Microsoft security response center - security bulletin severity rating
    system (2007),
    http://www.microsoft.com/technet/security/bulletin/rating.mspx,
33. Mell, P., et al.: Common vulnerability scoring system. IEEE Security and Privacy 4,
    85–89 (2006)
34. Rieke, R.: Modelling and analysing network security policies in a given vulnerability
    setting. Critical Information Infrastructures Security (2006)
35. Eschelbeck, G.: The laws of vulnerabilities: Which security vulnerabilities really
    matter. Information Security Technical Report (2005)
36. Chen, Y.: Stakeholder value driven threat modeling for off the shelf based systems
    (2007)
37. Liu, N., et al.: Security assessment for communication networks of power con-
    trol systems using attack graph and mcdm. IEEE Transactions on Power Delivery
    (2010)
38. Ni, M., et al.: Online risk-based security assessment. IEEE Transactions on Power
    Systems (2003)
39. Rezmierski, V., et al.: Incident cost analysis and modeling project (i-camp). Tech-
    nical Report, Higher Education Information Security Council, HEISC (2000)
40. Allied World Assurance, Tech404 Data Loss Cost Calculator (2013),
    http://www.tech-404.com/calculator.html
41. Anderson, R., et al.: Measuring the cost of cybercrime. In: Proc. of Workshop on
    the Economics of Information Security, WEIS (2012)
42. Detica and C. Office, The cost of cyber crime: joint government and industry
    report. In: Detica Report (2012),
    https://www.gov.uk/government/publications/
    the-cost-of-cyber-crime-joint-government-and-industry-report

Idea: Security Engineering Principles for Day Two Car2X Applications

Sibylle Fröschle[1] and Alexander Stühring[2]

[1] University of Oldenburg & OFFIS, Oldenburg, Germany
froeschle@informatik.uni-oldenburg.de
[2] University of Oldenburg, Oldenburg, Germany
alexander.stuehring@informatik.uni-oldenburg.de

Abstract. Car2X communication based on IEEE 802.11p has been prepared over the last years by a tremendous collaborative effort of the automotive industry, researchers, and standardization bodies. At the European level ETSI has released a first set of standards for interoperability and initial deployment in February this year. The "day one" Car2X applications, which will be implemented soon, do not automatically intervene into the driving behaviour of a car, but such "day two" applications will be the next step. In this idea paper we highlight some principle differences between Car2X and traditional security-relevant systems such as banking. We show how this might lead to a more straightforward escalation of attacks, which should be kept in mind in view of the safety-relevant "day two" applications. We also propose a principle of how to avoid this. In current work we test the feasibility of the attacks.

1 Introduction

Car2X communication based on IEEE 802.11p has been prepared over the last years by a tremendous collaborative effort of the automotive industry, researchers, and standardization bodies. At the European level ETSI[1] has released a first set of standards for interoperability and initial deployment in February this year.[2] These standards will enable the deployment of a set of "day one" Car2X applications, which will improve road safety and traffic efficiency. As an example consider the use case "Emergency electronic brake lights" [5]: if a vehicle performs an emergency brake then emergency messages will be broadcast to warn the follower vehicles of this event. The messaging is triggered by the braking vehicle when the switch of its emergency electronic brake lights is set on. Thereby the risk of longitudinal collision will be reduced.

"Day one" applications communicate Car2X messages as warnings or contextual awareness information to the driver, and do not automatically intervene into the driving behaviour of a vehicle. However, it is clear that Car2X will reach its

---

[1] European Telecommunications Standards Institute: http://www.etsi.org
[2] C.f. http://www.etsi.org/news-events/events/618-2013-itsworkshop

J. Jürjens, F. Piessens, and N. Bielova (Eds.): ESSoS 2014, LNCS 8364, pp. 213–221, 2014.
© Springer International Publishing Switzerland 2014

full potential only with actively intervening "day two" applications. For example, in the "Emergency electronic brake lights" use case, it is technically feasible to automatically brake follower vehicles.

It is immediately clear that if Car2X messages were sent unauthenticated an attacker could easily wreak havoc by injecting forged messages. In particular, with repect to "day two" applications such attacks could have a very drastic impact on the safety of road users. This and other security risks as well as privacy concerns have been taken into account: based on research projects such as EVITA[3], SeVeCom[4] and PRESERVE[5], the ETSI standards advocate a sophisticated security architecture, which includes authenticated Car2X communication, cryptographic keys and credentials management, privacy enabling technologies by pseudonyms, and in-car software and hardware security [13,8,6].

In this paper, we put forward ideas towards answering the following questions: what is the likelihood that something goes wrong with this security architecture, what is the safety impact, and how can the risk be contained? Our focus will be on the security of user credentials. After summarizing the ETSI security architecture we present our ideas: (1) By a comparison with the traditional security-relevant area of banking we highlight that Car2X user keys are employed under novel trust assumptions, which might lead to a more straightforward escalation of attacks. Thus motivated, we propose several robustness principles for security engineering Car2X. (2) We explore the scalability of attacks within the current ETSI architecture, using the "Emergency electronic brake lights" use case as a concrete example. We exhibit two scenarios that if realized would allow for an attack escalation, and show that our containment principle is not adhered to by the current Car2X architecture. In current practical work we are working towards a proof-of-concept of the attacks. (3) In view of the safety-relevant "day two" applications, we suggest a "security by design" method, which would contain the attacks.

2 The ETSI Communication and Security Architecture

We provide a summary of the ETSI ITS (Intelligent Transportation Systems) communication and security architecture as specified in [8,6].

To participate in the ITS every vehicle first has to enrol with an *enrolment authority*. As a result of enroling with the enrolment authority *ea* the vehicle *v* will possess a long-term signature key sk_v together with an *enrolment ticket* $ecert_{ea}(pk_v)$, a certificate for the corresponding public key pk_v signed by *ea*. With these long-term credentials the vehicle can then apply to an *authorization authority* to obtain authorization for a particular Car2X application or service. After successfully applying to the authorization authority *aa* the vehicle *v* will hold a fresh short-term signature key sk_p together with an *authorization ticket* $acert_{aa}(pk_p)$, a certificate

---

[3] E-safety Vehicle Intrusion Protected Applications: http://evita-project.org
[4] Secure Vehicular Communication: http://www.sevecom.org
[5] Preparing Secure Vehicle-to-X Communication Systems:
   http://www.preserve-project.eu

for the corresponding public key pk_p signed by aa. The vehicle may obtain many such pairs for different *pseudonyms p*, and exchange their use frequently. This will prevent that the vehicle can be tracked by its trace of Car2X messages, and thus ensures privacy. A pilot PKI to provide the necessary public key infrastructure is currently run by the Car2Car Communication Consortium [2].

For safety applications personal user vehicles will typically obtain authorization to broadcast the two types of basic Car2X messages: *cooperative awareness messages (CAMs)* and *distributed environmental notification messages (DENMs)*. CAMs are beaconed periodically by all vehicles to inform neighbour nodes about their position, speed, and direction. The idea is that based on CAMs every vehicle can maintain a *local dynamic map (LDM)* of their environment. In contrast, DENMs are triggered by an event such as an emergency brake. They are then disseminated into the relevant geographical area by multi-hop broadcasting. Depending on the use case DENMs have a typical life cycle. For example, in the "Emergency electronic brake lights" use case the originating vehicle will periodically broadcast DENMs until the emergency brake lights are off. A DENM cancellation message can be sent to signal the termination of the event [9].

To prevent message injection and tampering CAMs and DENMs are always sent authenticated. Each vehicle signs its outgoing CAMs and DENMs with its signature key sk_p and prepends the corresponding authorization ticket $cert_{aa}(pk_p)$. Thus, a receiving vehicle will be able to validate the signature. Authorization certificates and signed messages are implemented by the WAVE 1609.2 standard, which uses the relatively efficient ECDSA as signature algorithm [7].

3   Trust Assumptions and Robustness Principles

Having seen the security architecture of Car2X we now highlight some differences to traditional security-relevant networks such as the EMV system, which is used worldwide for smartcard banking transactions at point-of-sales or ATMs. Thus motivated, we derive several robustness principles for security engineering Car2X.

The security architectures of both banking and Car2X require that each user is equipped with authentication credentials: in banking each user is issued with a smartcard, which contains a signature key and a certificate for authentication with an ATM or sales terminal (e.g. [12]); in Car2X each car is equipped with a security module, which contains a signature key and a certificate for obtaining authorization for Car2X applications (and ultimately signature keys and certificates that authorize the car to send CAMs and DENMs). However, the "enabling scope" of such credentials is in each case very different: *While in banking the credentials of one user give access to the one user's resources, in Car2X the credentials of one user give access to influencing the driving behaviour of all peers at all locations within the authorization realm (which for CAMs and DENMs is at least nationwide).*

Hence, user credentials need to be well-protected, including protection from abuse by the car owner themselves. The ETSI security architecture prescribes that ITS key material should be stored in a tamper-resistant hardware security module (HSM). A good security API guarantees that sensitive keys are generated within the HSM, and that they will never be revealed in plaintext. (This is possible since all cryptographic operations can be performed within the HSM, accessing keys only via handles.) Good tamper-resistance ensures that the keys cannot easily be obtained by physical means. However, the current Car2X security architecture does not specify to which standard the security API of the HSM or its tamper-resistance should be evaluated. Neither do vendors of Car2X modules advertise any details to this point.[6]

By now it is well-known that security APIs are susceptible to subtle attacks where an attacker can trick an HSM into revealing a sensitive key by using the API in an unanticipated fashion [1,3]. Fortunately, there are now formal methods available that allow us to verify the API relative to standard assumptions similar to those in formal security protocol analysis [3,10]. Hence, we propose:

Principle 1 (Software security of user credentials). *To prevent that user credentials can be obtained by software attacks, the security APIs of Car2X HSMs should be formally verified.*

Concerning security against physical attacks (and also cryptographic attacks) one should keep a further difference to banking in mind: *While in banking user keys can easily be revoked in case of theft of the smartcard (and the owner themselves would only be able to access resources they already own), in Car2X an attacker can experiment with their own device for a long period of time, and then use the retrieved keys in an attack against others.* High-cost HSMs, evaluated to FIPS-140-2 level 4, provide a complete envelope of protection; they can detect physical penetration, and then zeroize sensitive memory immediately. However, it is highly unlikely that cost-effective mass-market HSMs such as those needed for Car2X will ever reach this level. For mass-market TPMs physical attacks are well-known.

While it seems clear that a dedicated attacker with large resources could extract a key it is difficult to estimate what their motive would be. If the key enabled them to perform attacks with a large safety impact then one would have to consider the threat of cyber war or cyber terrorism. One should also keep in mind that safety-relevant attacks, or rather the ability to execute them, could be used to blackmail or coerce OEMs and authorities: with an ultimately financial, political, or military motive. Since attacker types with large resources do potentially exist we recommend to adopt the following robustness principle:

Principle 2 (Physical security of user credentials). *The global safety of the Car2X system should not be based on the assumption that user credentials such as long-term vehicle signature keys cannot be obtained by an attacker who has large resources and physical access to the HSM that holds the keys.*

What could a global safety goal of a Car2X system be under this assumption? It is important to keep in mind that the current transportation system

---

[6] C.f. http://www.auto-talks.com/public/page.aspx?PageID=27
and https://www.escrypt.com/index.php?id=201

is very vulnerable to safety attacks. Anyone could at any time endanger others by dangerous driving or more drastic actions. As a very tragic example take an incident in Oldenburg, where a person threw a 6 kg block of wood from an autobahn bridge, thereby killing a passenger in a passing car.[7] This is by no means the only such incidence. Theoretically such attacks could be scaled to a large impact by a cell of terrorists performing instances of this attack at the same time at various locations. What keeps it in check is that the number of participating terrorists grows proportional to the scale of the attack. We translate this into the principle of proportion between attack distribution and physical presence.

Principle 3 (Proportion of attack distribution and physical presence). *The Car2X system should satisfy the principle of proportion of attack distribution and physical presence: whenever there is a distributed attack at locations $x_1, \ldots, x_n$ then n physical entities must be present at $x_1, \ldots, x_n$.*

4   How Much Damage Can You Do with One Set of User Credentials?

Let us now explore the scalability of attacks within the current ETSI architecture. Motivated by the previous discussion we assume:

Assumption 1. *The attacker group has obtained a set of Car2X user credentials: a long-term signature key sk_v with a corresponding enrolment certificate, as well as a signature key sk_p for signing DENMs and CAMs with a corresponding authorization certificate.*

To be able to argue concretely we consider the attacker group's ability to forge emergency brake light situations. An emergency brake light situation is forged when an honest vehicle decides to deliver an HMI warning to the driver (or, if it was "day two", to induce automatic braking) while no vehicle has actually performed an emergency brake at the supposed location.

We assume that to reduce any security risk vehicles use as much contextual awareness information as possible to validate received DENMs before acting on them. However, we assume a no-line-of-sight situation where the receiver cannot validate a DENM by comparison with its own sensor input (e.g. due to bad wheather conditions such as fog). Altogether, before acting on a DENM a vehicle will: (1) validate the signature of the DENM, (2) check whether the situational information within the DENM such as time and event position is plausible and relevant, (3) check whether the DENM fits in with a vehicle profile within the local dynamic map (LDM) (based on CAMs), (4) test whether the pattern of the received DENMs and CAMs matches the usual pattern of an emergency brake light situation. Note that (3) and (4) are in general made difficult by the fact that vehicles may change pseudonyms frequently but we wish to explore the scalability of attacks under the worst conditions for the attacker here.

---

[7] http://www.sueddeutsche.de/panorama/
urteil-gegen-holzklotz-werfer-lebenslang-wegen-mordes-1.450602

1. Local Attack: The attacker group forges one emergency brake light situation at location x with one attacker (and possibly a helper) physically present at x.

Since the attacker has a signature key sk_p and a corresponding authorization certificate for sending DENMs and CAMs she can easily construct DENMs and CAMs with situational information of her choice. Thereby checks (1) and (2) can be passed easily. To not be caught out by the consistency test with the LDM, prior to sending out DENMs, the attacker participates in the traffic like an honest Car2X participant, beaconing CAMs signed with the key sk_p.

Altogether, she can forge a pattern of CAMs and DENMs typical for an emergency brake manoever: at the intended location of the phantom emergency brake she broadcasts the appropriate number of DENMs. Moreover, to simulate braking and breakdown she offsets the location information in the DENMs and CAMs from her true position and then stops sending them altgother. Since an emergency brake occurs within a short time span this should be possible while she is actually driving away from the location. Otherwise she can always use a helper on the roadside, who takes over sending DENMs and CAMs when she is too far away from the phantom emergency brake location.

2. Multiplication Attack: The attacker group forges n emergency brake light situations at locations $x_1, \ldots, x_n$ with n attackers (possibly each with a helper) physically present at $x_1, \ldots, x_n$.

This can be done analogously to above: n attackers distributed at n locations perform n instances of the local attack. They can use the same key credentials, which can be exchanged beforehand (by person, email, phone, etc.). The timing of the attacks can be synchronized to make it more spooky.

Note that if neither check (3) nor check (4) is carried out then the attack can easily be performed by one attacker who remotely controls n 802.11p boxes hidden at the roadside at the n locations.

3. Escalation Attack via Malware on Head Units: The attacker group forges a large number of emergency brake light situations at a large number of locations without any physical entities present, using a malware distributed on automotive head units.

We additionally assume a software vulnerability in the head unit of a widely used car model. This will allow the attacker group to widely distribute a piece of malware onto Car2X vehicles. This is plausible since there will be many info-tainment applications that users can download themselves. The attack will work even if we assume that the safety-critical code of the head unit runs in a secure execution environment and cannot be meddled with by the malware. We only assume that the malware has access to the 802.11p network interface (which can also be used for infotainment applications).

The malware waits for an environmental trigger, such as a specific date and time, and then forges an emergency brake light situation. This can be done as follows. The malware contains the signature key sk_p and the corresponding certificate, and by using the GPS and clock of the vehicle (which are available to infotainment applications) it can construct DENMs and CAMs with appropriate situational information similarly to the local attack but dynamically. To pass check (3) it can send out CAMs early on, using the 802.11p interface of the host vehicle. One question is whether the LDM consistency check of the receiving vehicle could spot that there are two "overlapping" vehicles, and thereby recognize that one must be a phantom vehicle. (Recall that we only assume infotainment access, which means that the host vehicle will continue beaconing authentic CAMs.) However, this could probably be overcome by offsetting the forged CAMs a little pretending the path of a motorbike. If the local attack works without a roadside helper then the malware can fake an emergency brake analogously.

This will not work in every case. But given a wide distribution of the malware it would probably cause a large enough number of forged emergency brake light situations to make this a serious attack. Also note that if the lifetime of sk_p is shorter than the expected distribution time of the malware then the malware could run a small IRC client for post-compromise control (e.g. making use of a vehicle's cellular interface as in [4]). Thereby fresh authorization credentials could be distributed to the malware instances via a central attacker system, which could update the credentials easily without arousing suspicion by using the long-term key sk_v.

4. Escalation Attack Via Malware on Standard Devices: The attacker group forges a large number of emergency brake light situations at a large number of locations without any physical entities present, using a malware distributed on standard devices.

While it will be difficult enough to keep malware out of head units it will be impossible to keep malware from infecting common mobile devices such as smartphones, tablets, and laptops. So the question is: is it possible to run an attack similar to the previous one from an infected common mobile device onboard a vehicle? The main hindrance of doing so is that such common devices are equipped with standard 802.11abg hardware rather than 802.11p, which is dedicated to vehicular communications. However, it might still be possible.

Since 802.11p devices have been rarely available and are still very expensive compared to commercial off-the-shelf 802.11abg devices researchers interested in analysing the 802.11p channel properties have resorted to employing commercial off-the-shelf 802.11abg hardware for their research [15,14]. The 802.11p specification is directly derived from 802.11a [11], and it seems possible to make standard 802.11abg devices speak 802.11p by merely modifying, in a linux-type operating system, kernel modules such as the card driver and the linux wireless subsystem. Such manipulations could be installed by malware with root privilege. In practical work, we are currently working towards testing the feasibility of this attack. It is e.g. not clear whether the signal strength of 802.11a chips in laptops will be sufficient.

5   Discussion and Solutions

The last two hypothetical attacks show that the current Car2X system might well allow for potentially large scale attacks that do not satisfy the principle of proportion between distribution and physical presence. In view of the safety-relevant "day two" applications such attacks could potentially be large enough in impact to be relevant for cyber-terrorism or cyber-war. For "day one" applications one could "deflate" such attacks by socio-technical measures: if Car2X users perceive the warnings as "look out for this" rather than "there is definitely such a situation" no harm is done.

Unfortunately, we have seen that plausibility tests can be very limited since an attacker with a set of user credentials has a great repertoire of forging situational information, perhaps, even dynamically as in Attack 3 and 4. Of course, one could always think of another measure. For example, Attack 3 could be prevented if the compromised vehicle ran intruder detection (from a secure execution environment) that would spot the phantom vehicle, etc. However, for a robust and verifiably robust security architecture trust should be anchored in clear assumptions.

The most robust way to go forward seems to be to design locality and proof of physical presence into the Car2X system. To see what we mean first consider a road hazard warning application, where a road side unit (RSU) triggers the DENMs. Assume that the attacker has got hold of the signature key of the RSU and a corresponding authorization certificate. Then analogues of Attacks 3 and 4 are easily preventend if the authorization certificate contains the location of the *static* RSU and receiving vehicles check whether it is plausible that the DENMs stem from an RSU at this location. Attack 2 is then only possible within a very restricted area. The potential of attack scalability is caused by the large authorization realm of *mobile* vehicles.

For Car2X applications that entirely rely on cooperative information rather than local sensor input it would make the system much more robust if vehicles were required to produce witness certificates of their physical presence: RSUs could compare an optically observed identifer such as the number plate of the vehicle with the same type of identifier provided by the vehicle in a long-term certificate. If both match then the RSU will provide the vehicle with a certificate, say for its pseudonym p, that certifies that p has been physically observed at location x and time t. Such location stamps could contain Attacks 3 and 4 (while Attack 2 was only possible when obtaining n keys and the correponding cars). Privacy relative to other vehicles is still given.

We have addressed here only principal issues arising from the (in)security of vehicle keys. There are of course many other facets such as software integrity of the Car2X head-unit, and how to bootstrap it verifiably from the automotive HSMs (and ultimately the vehicle keys!). Moreover, Car2X PKI servers will be *very* safety-critical infrastructure. That none of the hypothetical attacks should be taken lightly should be clear latest since [4].

Acknowlegdments. The first author would like to thank the PRESERVE community for many discussions at the PRESERVE Summerschool and the CAR2X Architecture Workshop. This work was supported by the funding initiative *Niedersächsisches Vorab* of the Volkswagen Foundation and the Ministry of

Science and Culture of Lower Saxony (as part of the *Interdisciplinary Research Center on Critical Systems Engineering for Socio-Technical Systems*).

References

1. Anderson, R., Bond, M., Clulow, J., Skorobogatov, S.: Cryptographic processors—a survey. Proceedings of the IEEE 94, 357 (2006)
2. Bißmeyer, N., Stuebing, H., Schoch, E., Götz, S., Stotz, J.P., Lonc, B.: A generic public key infrastructure for securing car-to-x communication. In: 18th ITS World Congress, Orlando, USA (2011)
3. Bond, M., Focardi, R., Fröschle, S., Steel, G.: Analysis of security APIs (Dagstuhl Seminar 12482). Dagstuhl Reports 2(11), 155–168 (2012)
4. Checkoway, S., McCoy, D., Kantor, B., Anderson, D., Shacham, H., Savage, S., Koscher, K., Czeskis, A., Roesner, F., Kohno, T.: Comprehensive experimental analyses of automotive attack surfaces. In: Proceedings of the 20th USENIX Conference on Security, SEC 2011. USENIX Association (2011)
5. ETSI: TR 102 638 V1.1.1: ITS; vehicular communications; basic set of applications; definitions (June 2009)
6. ETSI: TS 102 731 V1.1.1: ITS; security; security services and architecture (September 2010)
7. ETSI: TS 102 867 V1.1.1: ITS; security; stage 3 mapping for IEEE 1609.2 (June 2012)
8. ETSI: TS 102 940 V1.1.1: ITS; security; ITS communications security architecture and security management (June 2012)
9. ETSI: TS 101 539-1 V1.1.1: ITS; v2x applications; part 1: Road hazard signalling (rhs) application requirements specification (August 2013)
10. Fröschle, S.: From Security Protocols to Security APIS: Foundations and Verification (To appear in the Information Security and Cryptography series of Springer)
11. Institute of Electrical and Electronics Engineers: IEEE Standard for Information Technology, Local and Metropolitan Area Networks, Specific Requirements, Part 11: Wireless LAN Medium Access Control (MAC) and Physical Layer (PHY) Specifications Amendment 6: Wireless Access in Vehicular Environments. IEEE Standard 802.11p-2010, pp. 1–51 (2010)
12. Murdoch, S.J., Drimer, S., Anderson, R., Bond, M.: Chip and PIN is broken. In: IEEE Symposium on Security and Privacy, pp. 433–446 (2010)
13. Papadimitratos, P., Buttyan, L., Holczer, T., Schoch, E.: Secure Vehicular Communication Systems: Design and Architecture. IEEE Communcations Magazine 46(11), 100–109 (2008)
14. Schumacher, H., Tchouankem, H., Nuckelt, J., Kurner, T., Zinchenko, T., Leschke, A., Wolf, L.: Vehicle-to-Vehicle IEEE 802.11p performance measurements at urban intersections. In: IEEE International Conference on Communications (ICC 2012), pp. 7131–7135 (2012)
15. Vandenberghe, W., Moerman, I., Demeester, P.: Approximation of the IEEE 802.11p standard using commercial off-the-shelf IEEE 802.11a hardware. In: 11th International Conference on ITS Telecommunications (ITST 2011), pp. 21–26 (2011)

Idea: Embedded Fault Injection Simulator on Smartcard

Maël Berthier[1], Julien Bringer[1], Hervé Chabanne[1,2], Thanh-Ha Le[1], Lionel Rivière[1,2], and Victor Servant[1]

[1] SAFRAN Morpho
[2] Télécom ParisTech
Identity & Security Alliance (The Morpho and Télécom ParisTech Research Center)

Abstract. Smartcard implementations are prone to perturbation attacks that consist in changing the normal behavior of components in order to create exploitable errors. Perturbation attacks could be realized by different means such as laser beams involving costly and complex injection platforms. In the context of black box or grey box evaluation, there is a strong necessity of identifying fault injection vulnerabilities in developed products. This is why we propose to integrate the injection mechanism straight into the smartcard project. The embedded fault simulator program is thus integrated with the chip software and its effects can be analyzed by side-channel observations, which is not the case with any existing fault simulators. In this paper, we present this new concept and its architectural design. We show then how to implement the simulator on a real smartcard product. Finally, to validate this approach, we study the functional and side-channel impact of fault injection on a standard algorithm provided by the host smartcard.

Keywords: Fault injection simulation, fault attack, smartcard, embedded secure software.

1 Introduction

The telecommunication, banking and identity industry provides more and more services handling sensitive and personal data. New institutional and consumer services, entertainment and payment means confirm the trend. To avoid privacy/confidentiality disclosure, careful attention must be granted to this class of component security: smartcards. While processing, the semi-conductor based device leaks some physical signals such as electromagnetic radiations, the power consumption or execution timings. This exposes the smartcard to side-channel analysis as these signals are related to ongoing instruction sequence. Memory read/write and buses activity are of particular interest for side-channel observations as they store or manipulate sensitive data and instructions.

Implementations are also prone to perturbation attacks that consist in changing the normal behavior of components in order to create exploitable errors. This

---

This work was partially funded by the French ANR project E-MATA HARI.

J. Jürjens, F. Piessens, and N. Bielova (Eds.): ESSoS 2014, LNCS 8364, pp. 222–229, 2014.

can be done because hardware logic gates rely on metal-based semi-conductors, which can be tampered by any external mean influencing the electrical potentials. Fault injection attacks which can be performed using laser beam or electromagnetic pulse, abruptly increase the energy level of the chip causing uncontrolled gates freeze or switches, hence the fault.

For secure smartcard based products, side-channel and fault attacks scenarios must be considered. Security evaluations are often performed in black box or grey box approach. Vulnerabilities could then be reported but without pinpointing where or what causes the weakness. Replaying attacks following the same setup and parameters brings no more information. However, with an embedded fault simulator into an actual smartcard, replayed attacks can be monitored by side channels from which we can take advantage to locate the vulnerabilities in the source code.

The remainder of the paper is organized as follows: in Section 2, we review the physical threats to smartcards. In Section 3, we confront existing fault simulation solutions and show their limitations regarding our needs. We then describe our new fault simulator concept in Section 4. Prior to conclusion, we show the primary results obtained with our first implementation in Section 5.

2   Physical Threats to Smartcards

2.1   Side Channel Analysis

Since 1996, Kocher *et al* have attempted to break Diffie-Helmann, RSA, DSS using timing attacks [1]. Simple Power Analysis (SPA) and Simple Electromagnetic Analysis (SEMA)[2,3] aim to exhibit information depending on processed data on plain reading of the associated trace. In contrast, a statistical analysis is performed to point out dependencies between manipulated sensitive values and traces. This can be realized with Differential Power Analysis (DPA)[4] such as Correlation Power Analysis (CPA)[5] or Mutual Information Analysis MIA)[6]. DPA can be performed on simulated traces before the chip tape-out [7]. The idea of countermeasures against side-channel leakages consists in reducing the dependency between physical signals and manipulated data or instructions.

2.2   Fault Injection Attacks

Physical injection causes faults to spread across the chip resulting in operational malfunction. This is the source of errors affecting code or data manipulation. By controlling the injection precision in time and space, errors can be exploited in order to skip instructions, flip bits or generate other unexpected behaviors. In [8], Skorobogatov was able to avoid normal memory writes/erasures by disrupting its operation with laser faults. Consequently, while sensitive data are stored in memory, impeding the security memory flush with a fault attack could lead to sensitive information leak. These kinds of flaws have to be taken into consideration in the development phase of a smartcard-based product. Existing solutions to guard against fault injection are based on detection, tolerance and protection concepts.

In order to ensure a high level of security regarding established standards, these attack paths must be considered and tested in laboratories. The success of a laser injection does not depend only on spatial and temporal parameters. The evaluator has to consider wavelength, beam diameter, exposure duration and intensity. Laser beam is still an effective means to lead attacks but generally involves costly and cumbersome injection platforms. Physical attacks could be supplemented with full software simulations but at the expense of accuracy. For smartcard product developers, which possess physical cards, simulation techniques come to complement and validate physical injections through refinement of attack scenario.

3   Fault Injection Simulation

We distinguish three simulation approaches: emulation, classical and high-level. Classical simulation considers hardware models while high-level simulation focuses on the code only.

Considering a set of test parameters and a model for injecting fault, simulation allows us to determine faulty outputs, undetected faults and the fault coverage. Thereby, we can predict the behavior of a design prior to its physical implementation. Simulation models consider the impact of injected faults such as logic evaluations at logic and binary level [9]. Even with a parallel approach, simulation remains time consuming as the time complexity depends on the number of logic gates and models to be considered.

In contrast, emulation involves hardware designed to operate in the same way as smartcards, allowing runtime attacks [10,11]. By synthesizing design descriptions onto FPGAs, it leverages hardware acceleration and outperforms classical fault simulation. Therefore, emulators are bounded to specific devices, or a device family at best. For simulation and emulation, two main techniques stands out, namely saboteur and mutant. A saboteur is an extra code or module while a mutant [12] consists of a pre-existing code or a module that has been modified. Using an external controller, they can inject expected faults at a precise location (in code or module).

Static high-level simulation approaches consist of analyzing potential vulnerabilities at the source code level. The aim is to automatically detect vulnerabilities in the context of complex fault model such as multiple fault injections. Nothing ensures that the chosen fault effect is actually achievable but this approach remains powerful for static analysis.

Classical and high-level *simulation* approaches allow very complex fault models and can consider overpowered attackers while the *emulations*, depending on a physical implementation, is somehow less flexible but closer to an effective smartcard behavior. But theses techniques rely on assumptions about targets and fault models. For instance, even if an emulator tends to operate identical to the final target, it is still physically different and thus, will respond in different ways under fault injections. In particular, none of them allows evaluators to perform side-channel observations and fault injections simultaneously. In the context of black box or grey box evaluation, there is a strong need to identify fault injection vulnerabilities in developed products. This is why we propose to integrate the injection mechanism straight into the smartcard project. Here, a project stands for the whole platform source code of a smartcard based product. The embedded fault simulator program is thus integrated to the chip software and its effects can be analyzed by real side-channel observations.

4 Embedded Injection Fault Simulator Concept

4.1 Concept

We have integrated a fault injection mechanism right into the smartcard providing control commands to setup attacks and retrieve relevant data. The embedded fault mechanism acts as a self-test program with a high priority level, granting access to critical registers, memories and execution flow of the smartcard. One main advantage is that the very same fault simulator code could be embedded into various smartcard models and families. Thus, developers have nothing specific to worry about while working on projects for the referred component.

The fault simulator code is integrated to a common smartcard development project such that interactions remain possible regardless of the running application. After compilation and debug sessions, the project is loaded onto the final smartcard. Thus, it uses the existing APDU communication standard for I/O transmissions [13]. The embedded fault simulator can be seen as an operating system (OS) service to perform fault injections. Post-compilation parameters such as function identifiers (IDs) are saved for later use as inputs for the simulator.

4.2 Advantages

Due to its operating range, different fault models can be obtained with the embedded fault simulator such as code alterations (by control flow disruption) or data modifications at the register level. Based on the byte skipping and thanks to a configurable fault width (the number of bytes to be skipped), the instruction skipping fault model, which is a realistic model [14], is reached.

With the embedded fault simulator there is no need to model any instruction or physical behavior to reach a specific fault model. Compared to emulation, the hardware used in our case is the final target, and thus, no bias is introduced by peripheral devices. Furthermore, in contrast to simulation or emulation, injecting fault with the embedded simulator does not rely on mechanisms that modify the code of modules (Mutant) or add new ones (Saboteur) and it does not use any external injection controller.

Hardware modules, for example, a hardware Data Encryption Standard (DES) implementation do not constitute direct targets. However, their input parameters are loaded into registers, which are set up according to software functionalities (key loads, counter increments, mode of operation). Thereby, the embedded fault simulator could disrupt software initialization, chaining or termination phases.

This approach has several advantages such that fault injections will occur directly on the real physical component running a real project. This new embedded approach allows side-channel observations, which is not the case with any existing fault injection simulation technique.

In the fault injection context, unpredicted fault effects reduce the side-channel simulation precision. Power and noise models rely on assumptions about its parameters and electromagnetic models are even more complex. As depicted in Figure 1, side-channel observations possible by our approach rely on real physical observations and traces contain all the device activity. This gives us the possibility to perform fault simulations and side-channel observations simultaneously.

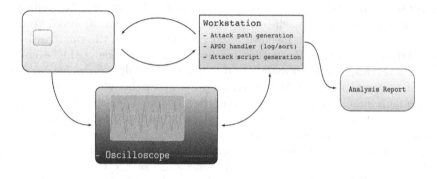

Fig. 1. Setup Functional Design

The simulator allows forward analysis, which consists of pinpointing fault vulnerabilities from code and confirms them with side-channel traces. But our main goal is to take advantage of side-channel signals to reveal software weaknesses, by analyzing patterns resulting from faulted execution traces.

In response to the lack of realism of simulation and the low success rate of physical injection, the idea is to bring the mechanism responsible for faults straight into the device under test. We keep the advantage of 100% injection success rate whereas laser fault injection could have no effect due to too weak beam intensity for instance. We also loose the spatial placement problem inherent to physical injections. But as stated before, the simulator could help refining the parameter range of physical injections.

4.3   Prototype Implementation

To demonstrate its feasibility, we have implemented a prototype on a real component and reached the instruction skipping fault model as a *proof-of-concept*. We studied the functional and the side-channel impact of fault injections on various functionalities provided by the host smartcard such as PIN verification, software security mechanisms or cryptographic software implementations.

The instruction skipping fault model consists of bypassing a determined instruction or a set of instructions while a program is running. Depending on the targeted instruction set, functional security calls or conditional tests can be avoided resulting in code re-branching. This leads to security breaches, which can potentially reveal information about the running code or manipulated data.

The high level code consists of a module interacting with the smartcard OS in order to provide the simulator program as service to all other embedded applications. On the other hand, the mechanism in charge of fault injections is defined at assembly level. The overhead resulting from the simulator integration represents no more than 1KB in its actual implementation, which is an overwhelming advantage in smartcard development.

As the fault occurs in the hardware abstraction layer, the fault simulator can impact all higher levels, from OS to application. The chosen method focuses on functions that are tagged with IDs, but is not restricted to this class of target. The fault simulator is embedded (and compiled) only in debug mode. There is no embedded fault simulator compiled in release versions or on the smartcard product to be shipped.

5 Fault Simulator Impact on Real Smartcards

5.1 Impact on Commands

For this study, we targeted a software implementation of the DES algorithm. First, a fault-free run is launched from which we store states for further comparison purposes. Those states consist of functional outputs and parameters from the fault simulator. Then, faults are injected during the whole DES execution and output behaviors are sorted out.

As shown in Table 1, eight behaviors were observed in our tests. Attack status is to be considered with a functional point of view, i.e. the target function terminates with or without the expected output. Functional failure simply means

Table 1. Encountered Functional Behavior Under Attack

Output	Behavior	Attack Status
Normal	Normal	–
Tampered	Altered but functional	Success
MuteCard	Altered and detected	Failure
KillCard	Altered and detected as critical	Failure
HardKill	Critical, Reflash required even in debug	Failure
Timeout	No answer after 10s	Failure
Freeze	Software card disconnect have no effect	Failure
Unknown	Unexpected values and behavior	?

no readable output. Therefore, the fault occurs in a specific time frame, which could be analyzed using side-channel observations in order to identify the program rerouting. In the context of fault injection, the failure case has a high probability to be reached. A `freeze` can occur mostly when the fault simulator creates infinite loops or leads to illegal values. The latter happens if an instruction is not included in the smartcard instruction set, consequently, the card switches to the freeze state.

5.2   Impact on Side Channel

The traces in Figure 2 depict the code bypassing effect obtained with the Embedded Fault Simulator. The light grey trace on top is with the embedded fault simulator running, the black trace is the reference trace without the fault simulator running. Before triggering the interruption (left part), the two traces match perfectly in time and amplitude. This is expected as nothing has yet occurred, but it shows the low footprint of the Embedded Fault Simulator.

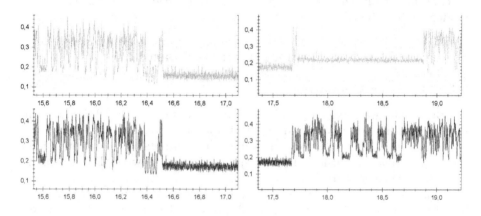

Fig. 2. Side Channel Impact - Targeted Functionality

In the light grey trace undergoing the attack, the simulator is set to perform a fault shortly after the entry point of a DES encryption. The right part depicts the impact of the fault simulator that shows de-synchronization and de-coherence between the two traces. A whole part is missing on the faulted trace (light grey). Retrieving the simulator parameters that led to this particular case allows us to point out the sensitive portion of the source code. It turned out that a conditional branch was bypassed preventing a while loop to call an external function. This results in an empty while loop that is observable on the faulted trace. For example, a fault bypassing the so-called 'try counter' during a PIN verification would allow brute force attacks.

6   Conclusion

In this paper we exposed a new concept for fault injection simulation directly integrated into the smartcard source code and we proposed a very compact implementation. We tested its capabilities on standard algorithms and reached the instruction skipping fault model. With the Embedded Fault Simulator we can reproduce behaviors obtained with laser injection. Moreover, this new approach allows side-channel observations from which we can take advantage to refine fault injection vulnerability detection in source code.

References

1. Kocher, P.C.: Timing Attacks on Implementations of Diffie-Hellman, RSA, DSS, and Other Systems. In: Koblitz, N. (ed.) CRYPTO 1996. LNCS, vol. 1109, pp. 104–113. Springer, Heidelberg (1996)
2. Gandolfi, K., Mourtel, C., Olivier, F.: Electromagnetic Analysis: Concrete Results. In: Koç, Ç.K., Naccache, D., Paar, C. (eds.) CHES 2001. LNCS, vol. 2162, pp. 251–261. Springer, Heidelberg (2001)
3. Sauvage, L., Danger, J., Guilley, S., Homma, N., Hayashi, Y.-I.: Advanced Analysis of Faults Injected Through Conducted Intentional Electromagnetic Interferences. IEEE Transactions on Electromagnetic Compatibility 55(3), 589–596 (2013)
4. Kocher, P.C., Jaffe, J., Jun, B.: Differential power analysis. In: Wiener, M. (ed.) CRYPTO 1999. LNCS, vol. 1666, pp. 388–397. Springer, Heidelberg (1999)
5. Coron, J.-S., Kocher, P.C., Naccache, D.: Statistics and Secret Leakage. In: Frankel, Y. (ed.) FC 2000. LNCS, vol. 1962, pp. 157–173. Springer, Heidelberg (2001)
6. Gierlichs, B., Batina, L., Tuyls, P., Preneel, B.: Mutual Information Analysis. In: Oswald, E., Rohatgi, P. (eds.) CHES 2008. LNCS, vol. 5154, pp. 426–442. Springer, Heidelberg (2008)
7. Hartog, J., Verschuren, J., Vink, E., Vos, J., Wiersma, W.: PINPAS: A Tool for Power Analysis of Smartcards. In: Security and Privacy in the Age of Uncertainty. IFIP, vol. 122, pp. 453–457. Springer, US (2003)
8. Skorobogatov, S.: Optical Fault Masking Attacks. In: FDTC, pp. 23–29. IEEE Computer Society (2010)
9. Berthomé, P., Heydemann, K., Kauffmann-Tourkestansky, X., Lalande, J.-F.: High Level Model of Control Flow Attacks for Smart Card Functional Security. In: ARES, pp. 224–229. IEEE Computer Society (2012)
10. Grinschgl, J., Aichinger, T., Krieg, A., Steger, C., Weiss, R., Bock, H., Haid, J.: Automatized Fault Attack Emulation for Penetration Testing. In: 12th International Common Criteria Conference (2011)
11. Kosuri, V.K., Fazal, N.: FPGA Modeling of Fault-Injection Attacks on Cryptographic Devices. IJERA 3, 937–943 (2013)
12. Machemie, J.-B., Mazin, C., Lanet, J.-L., Cartigny, J.: SmartCM a smart card fault injection simulator. In: WIFS, pp. 1–6. IEEE (2011)
13. ISO/IEC 7816-4 Identification cards – Integrated circuit cards – Part 4: Organization, security and commands for interchange(2013)
14. Moro, N., Dehbaoui, A., Heydemann, K., Robisson, B., Encrenaz, E.: Electromagnetic Fault Injection: Towards a Fault Model on a 32-bit Microcontroller. In: FDTC, pp. 77–88. IEEE (2013)

Author Index